Problem Solving Guide
and
Solutions Manual

to accompany

Russell

GENETICS

Problem Solving Guide
and
Solutions Manual

to accompany

Russell

GENETICS

Third Edition

Gail Patt
Boston University

■ HarperCollins*CollegePublishers*

Problem Solving Guide and Solutions Manual to accompany Russell's GENETICS, Third Edition

Copyright © 1992 by HarperCollins Colege Publishers Inc.

ISBN: 0-673-52201-6

92 93 94 95 96 9 8 7 6 5 4 3 2 1

TABLE OF CONTENTS

Preface

For many students, introductory genetics is their first, often difficult, encounter with analytical thinking and quantitative biology. As a result, this study guide/solutions manual has been designed to help students develop the skills necessary for the organization and translation of verbal information to a form that enables a quantitative solution.

Each chapter contains a short introduction to the peculiarities of solving the particular kinds of problems that are encountered in the chapter. There are a few multiple choice problems to set the wheels of thinking and remembering in motion, and these are followed by some questions that require more thoughtful answers or answers that require comparison and integration of several separate concepts. The solutions of the chapter-end problems follow. Many of these have expansive answers because the goal has been to try to substitute for a small class discussion with a good teacher. Often a student will know that a particular approach is the correct one, but not know why it is correct or why another approach is not. These are the kinds of questions that arise naturally in a discussion class, and it is these kinds of questions that have been anticipated here.

Two general suggestions must be made. First, there is no substitute for thoughtful reading. The text must be carefully read and its information must be manipulated. This manual should be considered to be a set of helpful guides, but it is not a substitute for the textbook. In addition, the chapter-end questions must be read carefully. Many mistakes are made in solution because of careless or hasty reading of the questions. Second, neither time nor experience can be compressed. Skills require practice. The more problems that are attempted, the more that will be solved, and the greater the level of analytical skill achieved. Some students seem to be born with great problem-solving abilities. For most, however, analysis is a learned skill. It can be mastered, but to do it well requires practice, and practice requires time.

I wish to express my thanks to Peter Russell, whose textbook embodies a model of clear and sympathetic writing with the student always in mind. Kathleen Dolan has been as willing and positive a development editor as could be wished, and Donald I. Patt has contributed countless hours of assistance in the preparation of all phases of this guide.

September 1, 1991
Princeton, Massachusetts

CHAPTER 1

VIRUSES, CELLS, AND CELLULAR REPRODUCTION

I. CHAPTER OUTLINE

Viruses
Cells - The Units of Life
    Prokaryotes
    Eukaryotes
Cellular Reproduction in Eukaryotes
    Chromosome Complement of Eukaryotes
    Asexual and Sexual Reproduction
    Mitosis: Nuclear Division
    Meiosis
    Genetic Significance of Meiosis
    Locations of Meiosis in the Life Cycle

II. IMPORTANT TERMS AND CONCEPTS

genetics	sister chromatids
gene	daughter chromosomes
mutation	pro-, meta-, ana-, telophase
recombination	tubulin
chromosome	aster
selection	spindle
recombinant DNA	metaphase plate
biotechnology/genetic engineering	gametogenesis
virus	meiospore
DNA	sporogenesis
RNA	disjunction
cell	meiosis I *vs.* II
microscopy	leptonema
prokaryote *vs.* eukaryote	zygonema
ribosome	pachynema
cyanobacteria	diplonema
nucleus	synapsis
plasma membrane	bivalent
nuclear membrane	tetrad
annulus	crossing-over
nucleolus	synaptonemal complex
cytoplasm	chiasma
cytoskeleton	diakinesis
mitochondrion	spermatogenesis
chloroplast	oogenesis
centriole	primary *vs.* secondary spermatogonium
endoplasmic reticulum	primary *vs.* secondary spermatocyte
Golgi apparatus	primary *vs.* secondary oogonium

microtubule
centrosome
chlorophyll
photosynthesis
gamete *vs.* zygote
haploid *vs.* diploid
homologous *vs.* non-homologous
sex chromosome *vs.* autosome
centromere *vs.* kinetochore
metacentric *vs.* submetacentric
telocentric *vs.* acrocentric
sexual *vs.* asexual reproduction
vegetative reproduction
somatic cell *vs.* germ cell
mitosis *vs.* meiosis
cyto- *vs.* karyokinesis
cell cycle
interphase
G_1 *vs.* G_2
S phase
chromatid

primary *vs.* secondary oocyte
spermatid *vs.* ootid
ovum
polar body
gametophyte
sporophyte
stamen *vs.* carpel
anther
pollen grain
stigma *vs.* style
ovary *vs.* ovule
seed
micro *vs.* megaspore mother cell
micro *vs.* megaspore
endosperm
embryo sac
generative nucleus
tube nucleus
micropyle
double fertilization
alternation of generations

III. THINKING ANALYTICALLY

Although the material in Chapter 1 is primarily descriptive there are several key points to remember when answering the chapter end questions.

1. The alignment of bivalents on the metaphase plate of meiotic metaphase I is random. Hence, it results in gametes that exhibit all possible combinations of maternally and paternally derived chromosomes.

2. Crossing-over is believed to be a random event as well. As a result, the chromosomes that exist after meiotic prophase I are different from those that entered meiosis.

3. These random events produce a range of chromosomally organized variability that adds to the more basic variations created by mutation and chromosomal aberration. Taken together, all of these sources of genetic change are vital at all levels of genetics - the molecular, the organismal, and the evolutionary.

4. Random processes generally require statistical solutions. Keep in mind the product rule for coincident events (the probability of two independent events occurring together equals the product of their individual probabilities) and the sum rule (the probability of one of several mutually exclusive events is the sum of their individual probabilities).

IV. QUESTIONS FOR PRACTICE

A. Multiple Choice Questions

1. If gametes are produced by meiosis in an animal species that has a diploid number 2N = 12, how many chromatids will be present in the primary oocyte?
   a. 3
   b. 6
   c. 12
   d. 24

2. In that same species, how many chromatids will be present in the second polar body?
   a. 3
   b. 6
   c. 12
   d. 24

3. In which of the following may DNA be single-stranded?
   a. viruses
   b. bacteria
   c. eukaryotes
   d. viruses, bacteria, and eukaryotes

4. In which of the following is DNA circular?
   a. bacteria
   b. lower eukaryotes
   c. higher eukaryotes
   d. bacteria and all eukaryotes

5. In which of the following do chromosomes consist of 2 chromatids during some stage of the life cycle?
   a. viruses
   b. bacteria
   c. eukaryotes
   d. viruses, bacteria, and eukaryotes

6. Which of the following cells enters meiosis in animals?
   a. oogonium
   b. primary spermatocyte
   c. secondary oocyte
   d. polar body

7. Which of the following cells enters meiosis in plants?
   a. megaspore mother cell
   b. microspore
   c. megaspore
   d. generative cell

Answers: 1d, 2b, 3a, 4a, 5c, 6b, 7a

B. Thought Questions

1. Are gametes always produced directly by meiosis? Must meiosis always be part of the sexual life cycle of a diploid organism? Hint: Review animal and plant life cycles.
2. Why are the sex chromosomes which, after all, have a different morphology, still considered to be "an homologous pair"? Hint: Review the role of homologous chromosomes in meiosis.
3. What are the similarities and differences between an oogonium and a macrospore mother cell?
4. Why do you think that gametophyte and sporophyte elements of a plant life cycle are designated as an "alternation of generations" whereas an animal's body and its gamete are not? Hint: Trace the fate of the meiotic product.

V. ANSWERS AND SOLUTIONS TO TEXT QUESTONS

1.1 Interphase is a period corresponding to the cell cycle phases of
   a. mitosis.
   b. S.
   c. $G_1 + S + G_2$.
   d. $G_1 + S + G_2 + M$.

Answer: c

3

1.2.   Chromatids joined together by a centromere are called
        a. sister chromatids.
        b. homologs.
        c. alleles.
        d. bivalents (tetrads).

Answer: a

1.3.   Mitosis and meiosis always differ in regard to the presence of
        a. chromatids.
        b. homologs.
        c. bivalents.
        d. centromeres.
        e. spindles.

Answer: c

1.4   State whether each of the following statements is true or false. Explain your choice.
        a. The chromosomes in a somatic cell of any organism are all morphologically alike.
        b. During mitosis the chromosomes divide and the resulting sister chromatids separate at anaphase, ending up in two nuclei, each of which has the same number of chromosomes as the parental cell.
        c   At zygonema, any chromosome may synapse with any other chromosome in the same cell.

Answer: a. False. While this may be true for a particular organism or organisms, in general the chromosomes vary morphologically in size and centromere position.
        b. True.
        c. False. Only homologous chromosomes can synapse at zygonema.

1.5   Decide whether the answer to these statements is *yes* or *no*. Then explain the reasons for your decision.
        a. Can meiosis occur in haploid species?
        b. Can meiosis occur in a haploid individual?

Answer: a Yes, if a sexual mating system exists in that species. In that case, two haploid cells can fuse to produce a diploid cell which can then go through meiosis to produce haploid progeny. The fungi *Neurospora crassa* and *Saccharomyces cerevisiae* exemplify this positioning of meiosis in the life cycle.
        b. No, because a diploid cell cannot be formed and meiosis only occurs starting with a diploid cell.

1.6   The general life cycle of a eukaryotic organism has the sequence
        a. 1N → meiosis → 2N → fertilization → 1N.
        b. 2N → meiosis → 1N → fertilization → 2N.
        c. 1N → mitosis → 2N → fertilization → 1N
        d. 2N → mitosis → 1N → fertilization → 2N.

Answer: b

1.7.   Which statement is true?
        a. Gametes are 2N; zygotes are 1N.
        b. Gametes and zygotes are 2N.
        c. The number of chromosomes can be the same in gamete cells and in somatic cells.
        d. The zygotic and the somatic chromosome numbers cannot be the same.
        e. Haploid organisms have haploid zygotes.

Answer: c

1.8.   All of the following happen in prophase I of meiosis *except*
        a.   chromosome condensation.
        b.   pairing of homologs.
        c.   chiasma formation.
        d.   terminalization.
        e.   segregation.

Answer: e

1.9.   Give the names of the stages of mitosis or meiosis at which the following events occur:
        a.   Chromosomes are located in a plane at the center of the spindle.
        b.   The chromosomes move away from the spindle equator to the poles.

Answer: a. metaphase
                 b. anaphase

1.10   Given the diploid, meiotic mother cell in the following figure, diagram the chromosomes as they would appear
        a.   in late pachynema;
        b.   in a nucleus at prophase of the second meiotic division;
        c.   in the polar body resulting from oogenesis in an animal.

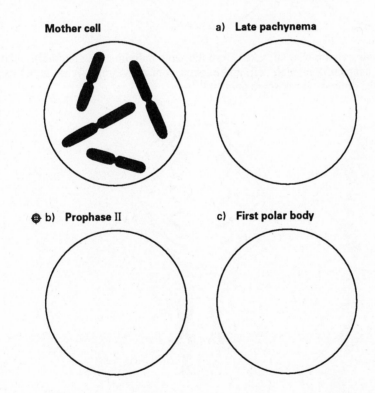

5

Answer:

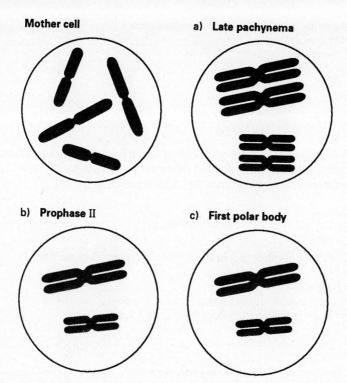

Mother cell

a) Late pachynema

b) Prophase II

c) First polar body

1.11. The cells in the following figure were all taken from the same individual (a mammal). Identify the cell division events happening in each cell, and explain your reasoning. What is the sex of the individual? What is the diploid chromosome number?

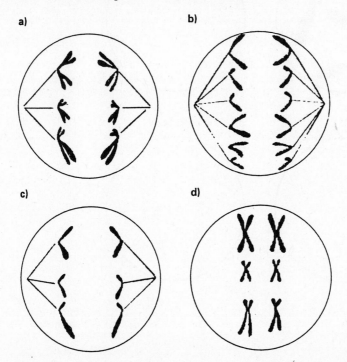

a)

b)

c)

d)

Answer: Cell a is in anaphase of meiosis I, because homologs are undergoing disjunction. Cell b is in anaphase of mitosis, because chromatids are disjoining, and there are 2N chromosomes (deduced by comparison to cell a). Cell c is in anaphase of meiosis II, because chromatids are disjoining and there are only N chromosomes. Cell d is in metaphase of meiosis I, as

6

homologs are paired. The individual is probably female, since there is no heteromorphic chromosome pair. The diploid chromosome number is three.

1.12   Does mitosis or meiosis have greater significance in genetics (i.e. the study of heredity)? Explain your choice.

Answer: Meiosis has greater significance. The role of mitosis is to produce progeny cells that are genetically identical to the parental cell. In contrast, meiosis generates great genetic diversity through the various ways in which the maternal and paternal chromosomes are combined in the progeny nuclei and by crossing-over events that produce nonparental combinations of genes in the resulting meiotic products.

1.13   Consider a diploid organism that has three pairs of chromosomes. Assume that the organism receives chromosomes A, B, and C from the female parent and A', B', and C' from the male parent. To answer the following questions, assume that no crossing-over occurs.
    a.   What proportion of this organism would be expected to contain all the chromosomes of maternal origin?
    b.   What proportion of the gametes would be expected to contain some chromosomes from both maternal and paternal origin?

Answer: a.   1/8. The probability of a gamete receiving A is 1/2 since the choice is A or A'. Similarly the probability of receiving B is 1/2 and of receiving C is 1/2. Therefore, the probability of receiving A, B, and C is 1/2 x 1/2 x 1/2 = 1/8 (like the probability of tossing three heads in a row, the individual probabilities are multiplied together).
    b.   6/8 (= 3/4). The probability of gametes containing chromosomes from both maternal and paternal origin is 1 minus the probability that the gametes will contain either all maternal or all paternal chromosomes. The probability of a gamete containing all maternal chromosomes is 1/8 (part a) and that of a gamete containing all paternal chromosomes is also 1/8. Therefore the answer is 1 - 1/8 - 1/8 = 6/8, or 3/4.

1.14   Normal diploid cells of a theoretical mammal are examined cytologically at the mitotic metaphase stage for their chromosome complement. One short chromosome, two medium-length chromosomes and three long chromosomes are present. Explain how the cells might have such a set of chromosomes.

Answer: One of the long chromosomes and the short chromosome might be members of heteromorphic pair - i.e., X- and Y-chromosomes, respectively.

1.15.   Is the following statement true or false? Explain your decision. "Meiotic chromosomes may be seen after appropriate staining in nuclei from rapidly-dividing skin cells."

Answer: False. Skin cells, being somatic cells, divide by mitosis, hence no meiotic chromosomes.

1.16.   Is the following statement true or false? Explain your decision. "All of the sperm from one human male are genetically identical."

Answer: False. Owing to the randomness of independent assortment and to crossing-over both of which characterize meiosis, the probability of any two sperm cells being genetically identical is extremely remote.

1.17.   The horse has a diploid set of 64 chromosomes, and a donkey has a diploid set of 62 chromosomes- Mules (viable but usually sterile progeny) are produced when a male donkey is mated to a female horse. How many chromosomes will a mule cell contain?

Answer: The cell of the mule will contain 63 chromosomes. That is, the sperm of the donkey will carry the haploid number of chromosomes for it species, which is 62/2 = 31. The egg of the mare will carry the haploid number for its species, which is 64/2 = 32. Fusion of the 31-chromosome sperm with the 32-chromosome egg will give a cell with 63 chromosomes.

7

CHAPTER 2

MENDELIAN GENETICS

I. CHAPTER OUTLINE

Genotype and Phenotype
Mendel's Experiments
    Mendel's Experimental Design
Monohybrid Crosses and the Principle of Segregation
    The Principle of Segregation
    Representing Crosses with a Branch Diagram
    Confirming the Principle of Segregation: The Use of a Testcross
Dihybrid Crosses and the Mendelian Principle of Independent Assortment
    Branch Diagram of a Dihybrid Cross
    Trihybrid Crosses
"Rediscovery" of Mendel's Principles
Mendelian Genetics in Human
    Pedigree Analysis
    Examples of Human Genetic Traits

II. IMPORTANT TERMS AND CONCEPTS

hereditary Trait	dihybrid
gene	locus (gene locus)
genotype	dominant
phenotype	recessive
self-Fertilization	testcross
cross-Fertilization	Punnett square
pure-Breeding Strain	branch diagram
P Generation	probability
F_1, F_2, F_3 Generation	product rule
reciprocal cross	sum rule
allele (allelomorph)	pedigree
gamete	pedigree analysis
homozygous	genetic counseling
heterozygous	X-Linkage
hybrid	penetrance
monohybrid	expressivity

III. THINKING ANALYTICALLY

1. Read the problem through without pausing, in order to get the sense of it.
2. Read it through again, more critically, point-by-point, assigning descriptive gene symbols where you can appropriately. Use a dash tentatively where you're not sure.

3. Look for clues in the problem statement, progressing from the start to the finish, and from the finish to the start.

4. Check back to make sure that each part of your solution is in agreement with all the information given.

EXAMPLE:

Question: A black, long-haired guinea pig is crossed with a white, short-haired guinea pig. The incredibly large litter born consists of 8 black, short-haired and 9 white, short-haired babies. The reciprocal cross gave similar results: 7 black short and 6 white short. When the two black long-haired parents were mated, their offspring consisted of 7 black long and 2 white long individuals. Assume independent assortment. What is the genotype of the parents?

Solution: 1. Tentatively assign the symbols *B* and *b* to the alleles for color, and *L* and *l* for the hair length alleles.

2. The third cross gave close to a 3 black:1 white ratio, which suggests that the black parents are heterozygous and black is dominant. Therefore their offspring are tentatively *Bb* - - and *bb* - -. All the offspring of this cross were long-haired, and so these same parents, having identical genotypes, must be homozygous for the hair length gene. Therefore the genotype for the black long parents is either *Bb LL* or *Bb ll*.

3. The offspring of the first and second crosses were all short-haired and so both were homozygous for the hair length gene: one homozygous for the long-hair allele, and the other homozygous for the short-hair allele. This interpretation agrees with the conclusion reached above, but does not tell which of the hair length alleles is dominant. What one additional cross would resolve this question, assuming that you are limited to using only the animals described above?

4. Obviously a testcross cannot be made, because this requires the use of a homozygous recessive, and the data do not reveal which of the hair length alleles is recessive. Since both parents are homozygous at this gene locus, one for the long allele and one for the short allele, their offspring must be heterozygous. If they were crossed, their offspring would show a 3:1 ratio for hair length. This is the one remaining cross that would have to be made. If the result came out 3 short to 1 long, we would conclude that the short allele is dominant, and sticking to convention, we would designate this allele by the symbol *S*, and the recessive allele by *s*, thus replacing *L* and *l*. Our final solution to the problem would state that the original parents (the P generation) would have the genotypes *Bb ss* and *bb SS*.

5. The last step is to review the problem from the beginning in order to verify that the conclusion agrees throughout.

IV. QUESTIONS FOR PRACTICE

A. Multiple Choice Questions

1. A dominant gene is one that
    a. suppresses the expression of genes at all loci.
    b. masks the expression of neighboring gene loci.
    c. masks the expression of its allele.
    d. masks the expression of all the foregoing.

2. A dihybrid is an individual that
    a. is heterozygous throughout its genotype.
    b. is heterozygous for two genes under study.
    c. is the result of a testcross.
    d. is used for a testcross.

3. The cross of an uncertain genotype with a homozygous recessive genotype at the same locus is a
    a. pure-breeding strain.
    b. monohybrid cross.
    c. testcross.
    d. dihybrid cross.

4. The genotypic ratio of the progeny of a monohybrid cross is typically
   a. 1:2:1.
   b. 9:3:3:1.
   c. 27:9:9:9:3:3:3:1.
   d. 3:1.

5. The phenotypic ratio of a dihybrid cross involving dominance and recessiveness is typically
   a. 9:1.
   b. 1:2:1.
   c. 3:1.
   d. 9:3:3:1.

Answers: 1c, 2b, 3c, 4a, 5c

B. Thought Questions

1. Correlate the principles of Mendel with the events of meiosis.
2. Why did Mendel's discoveries remain "undiscovered" for so many years? Hint: When was the significance of chromosomes discovered?
3. Over a period of many years, many attempts were made to try to understand how physical traits are passed on from one generation to the next. Gregor Mendel was the first to make the breakthrough. How do you account for his success, in light of years of failure before him? Hint: Give a new twist to "the best laid plans o' mice an' men gang aft agley" (Robert Burns) by substituting "poorly" for "best".
4. As you will see in subsequent chapters, most genes code the synthesis of polypeptides or proteins, and many, if not most of these are enzymes of one kind or another. Enzymes catalyze biochemical reactions, including those involved in biosynthesis. Some such reactions require relatively large quantities of enzyme, whereas others require very little, but most require some. On the basis of this information, propose an explanation for the phenomena of dominance, recessiveness, and, if you dare, penetrance and expressivity. Hint: Sometimes recessiveness is due to failure of an allele to produce any enzyme, or a less active form of enzyme. The actions of other genes may modify the expression of a given gene. Think of an organism as an incredible mass of chemical substances, including perhaps millions of different enzymes, some of which compete with other enzymes for substrate.

V. ANSWERS AND SOLUTIONS TO TEXT QUESTIONS

2.1 In tomatoes, red fruit color in dominant to yellow. Suppose a tomato plant homozygous for red is crossed with one homozygous for yellow. Determine the appearance of (a) the F_1; (b) the F_2; (c) the offspring of a cross of the F_1 back to the red parent; (d) the offspring of a cross of the F_1 back to the yellow parent.

Answers: a. red
              b. 3 red, 1 yellow
              c. all red
              d. 1/2 red, 1/2 yellow

2.2 A red-fruited tomato plant, when crossed with a yellow-fruited one, produces progeny about half of which are red-fruited and half of which are yellow-fruited. What are the genotypes of the parents?

Solution: 1. The one-to-one ratio of progeny suggests the results of a testcross.
              2. The preceding problem states that red-fruitedness in tomatoes is dominant to yellow-fruited.
              3. Since the recessive trait appears in the progeny, its allele must have been present in the red-fruited parent of this cross. Therefore, the red-fruited parent must be heterozygous, and the yellow-fruited must be homozygous recessive.

Answer: *Rr* and *rr* are the genotypes of the parents.

10

2.3. In maize, a dominant gene *A* is necessary for seed color as opposed to colorless (*a*). A recessive gene *wx* results in waxy starch as opposed to normal starch (*Wx*). The two genes segregate independently. Give phenotypes and relative frequencies for offspring resulting when a plant of genetic constitution *Aa WxWx* is testcrossed.

Answer: The cross is *Aa WxWx* x *aa wxwx*. Considering the *A* gene, half of the progeny will be *Aa* and half will be *aa*. Considering the *Wx* gene, all progeny will be *Wxwx*. Combining the results one-half of the progeny will be *Aa Wxwx*, that is, colored seeds, normal starch, and the other half will be *aa Wxwx*, that is, colorless seeds, normal starch.

2.4 F_2 plants segregate 3/4 colored:1/4 colorless. If a colored plant is picked at random and selfed, what is the probability that more than one type will segregate among a large number of its progeny?

Answer: The F_2 genotypic ratio (if *C* is colored, *c* is colorless) is 1/4 *CC*:1/2 *Cc*:1/4 *cc*. If we consider just the colored plants, there is a 1:2 ratio of *CC* homozygotes to *Cc* heterozygotes. Therefore, if a colored plant is picked at random, the probability that it is *CC* is 1/3 (i.e., 1/3 of the *colored* plants are homozygous) and the probability that it is *Cc* is 2/3. Only if a *Cc* plant is selfed will more than one phenotypic class be found among its progeny; therefore, the answer is 2/3.

2.5 In guinea pigs rough coat (*R*) is dominant over smooth coat (*r*). A rough-coated guinea pig is bred to a smooth one, giving eight rough and seven smooth progeny in the F_1.
    a. What are the genotypes of the parents and their offspring?
    b. If one of the rough F_1 animals is mated to its rough parent, what progeny would you expect?

Answer: a. Parents are *Rr* (rough) and *rr* (smooth); F_1 are *Rr* (rough) and *rr* (smooth).
    b. *Rr* x *Rr* → 3/4 rough, 1/4 smooth

2.6 In cattle the polled (hornless) condition (*P*) is dominant over the horned (*p*) phenotype. A particular polled bull is bred to three cows. With cow A, which is horned, a horned calf is produced; with a polled cow B a horned calf is produced; and with horned cow C a polled calf is produced. What are the genotypes of the bull and the three cows, and what phenotypic ratios do you expect in the offspring of these three matings?

Answer:: Bull is *PP* or *Pp*. With A, a horned calf *pp* is produced, so bull must be *Pp*. Cow A has to be *pp*, so there should be 1/2 polled and 1/2 horned in the progeny. Cow B is polled and gives a horned calf, so the cow is *Pp*. Offspring from this cross are 3/4 polled and 1/4 horned. Cow C is horned and must be *pp*, and the polled calf produced is *Pp*. Offspring are 1/2 polled and 1/2 horned.

2.7 In the Jimsonweed purple flowers are dominant to white. When a particular purple-flowered Jimsonweed is self-fertilized, there are 28 purple-flowered and 10 white-flowered progeny. What proportion of the purple-flowered progeny will breed true?

Answer: Progeny ratio approximates 3:1, so the parent is heterozygous. Of the dominant progeny there is a 1:2 ratio of homozygous to heterozygous, so 1/3 will breed true.

2.8 Two black female mice are crossed with the same brown male. In a number of litters female X produced 9 blacks and 7 browns and female Y produced 14 blacks. What is the mode of inheritance of black and brown coat color in mice? What are the genotypes of the parents?

Answer: Black is dominant to brown. If *B* is the allele for black and *b* for brown, then female X is *Bb* and female Y is *BB*. The male is *bb*.

2.9 Bean plants may differ in their symptoms when infected with a virus. Some show local lesions that do not seriously harm the plant. Other plants show general systemic infection. The following genetic analysis was made:

P   local lesions X systemic lesions

F_1 all local lesions

F_2 785 local lesions:269 systemic lesions

What is probably the genetic basis of this difference in beans? Assign gene symbols to all the genotypes occurring in the above experiment. Design a test cross to verify your assumptions.

Solution:

1. Start by estimating what the numerical ratio is by adding the numbers of the two F_2 classes (785 + 269 = 1054); then calculate three quarters of the sum (0.75 x 1054 = 790.5) and one quarter (0.25 x 1054 = 263.5). These quantities are very, very close to the numbers given, and so it's safe to assume a 3:1 ratio by inspection. This ratio is typical of a monohybrid cross with complete dominance of one allele.
2. Assign symbol *A* for the dominant allele, *a* for the recessive allele.
3. The cross can be symbolized now by genotypes:

P   *AA* x *aa*
                       ↓

F_1 *Aa*
                       ↓

F_2 1 *AA*:2 *Aa*:1 *aa* = 3 *A*–:1 *aa*

4. Testcross: F_1 x *aa* → 1 *Aa*::1 *aa*

2.10 Fur color in babbits, a furry little animal and popular pet, is determined by a pair of alleles, *B* and *b*. *BB* and *Bb* babbits are black, and *bb* babbits are white.

    A farmer wants to breed babbits for sale. Pure *bb* female babbits breed poorly, so he mates a pair of black babbits, which produce six black and two white offspring. The farmer immediately sells his white babbits, then he comes to consult you for a breeding strategy to produce more white babbits.

   a   If he performed random crosses between pairs of F_1 babbits, what proportion of the F_2 progeny would be white?
   b. If he crossed an F_1 male to the parental female, what is the probability that this cross will produce white progeny?
   c. What would be the farmer's best strategy to maximize the production of white babbits?

Answer:

   a. The black babbits of cross (a), above, could be either *BB* or *Bb*, but you can't tell which is which by looking at them. The following combinations of matings could occur: *BB* x *BB*, *BB* x *Bb*, and *Bb* x *Bb*. The last is the only one that could produce white offspring, and as we know by now, the proportion of heterozygotes resulting from a monohybrid cross that show the dominant trait is 2/3. Accordingly, the probability of selecting two heterozygotes by chance alone is 2/3 x 2/3, and the probability that this mating will produce a white offspring is 1/4. Therefore the probability that the random matings among members of this group of black babbits = 2/3 x 2/3 x 1/4 = 4/36 = 1/9.
   b. The only way available to him to produce a white offspring is to pick a heterozygous (*Bb*) male, the probability of which is 2/3, and mating it to the parental female, which we already know is *Bb*. The probability of getting a white babbit from this cross is 2/3 x 1 x 1/4 = 1/6.
   c. His best strategy, then, would be to backcross a white male obtained from this cross to the parental females; thereby getting a 50% yield of white progeny.

2.11 In Jimsonweed purple flower (*P*) is dominant to white (*p*), and spiny pods (*S*) are dominant to smooth (*s*). In a cross between a Jimsonweed homozygous for white flowers and spiny pods and one homozygous for purple flowers and smooth pods, determine the phenotype of (a) the F₁; (b) the F₂; (c) the progeny of a cross of the F₁ back to the white, spiny parent; (d) the progeny of a cross of the F₁ back to the purple, smooth parent.

Answer: a. *Pp Ss;*so purple, spiny
           b. 9/16 purple, spiny:3/16 purple, smooth:3/16 white, spiny:1/16 white, smooth
           c. 1/2 purple, spiny:1/2 white, spiny
           d. 1/2 purple, spiny:1/2 purple, smooth

2.12   What progeny would you expect from the following Jimsonweed crosses? You are encouraged to use the branch diagram approach.
        a. *PP ss* x *pp SS*
        b. *Pp SS* x *pp ss*
        c. *Pp Ss* x *Pp SS*
        d. *Pp Ss* x *Pp Ss*
        e. *Pp Ss* x *Pp ss*
        f. *Pp Ss* x *pp ss*

Answer: a. all *Pp Ss*, purple, spiny
           b. 1/2 purple, spiny *(Pp Ss)* :1/2 white, spiny *(pp Ss)*
           c. 3/4 purple, spiny:1/4 white, spiny
           d. 9/16 purple, spiny:3/16 purple, smooth:3/16 white, spiny:1/16, white, smooth
           e. 3/8 purple, spiny:3/8 purple, smooth:1/8 white, spiny:1/8 white, smooth
           f. 1/4 purple, spiny:1/4 purple, smooth:1/4 white, spiny:1/4 white, smooth

2.13   In summer squash white fruit (*W*) is dominant over yellow (*w*), and disk-shaped fruit (*D*) is dominant over sphere-shaped fruit (*d*). In the following problems the appearances of the parents and their progeny are given. Determine the genotypes of the parents in each case.
        a. White, disk x yellow, sphere gives 1/2 white, disk and 1/2 white, sphere.
        b. White, sphere x white, sphere gives 3/4 white, sphere and 1/4 yellow, sphere.
        c. Yellow, disk x white, sphere gives all white, disk progeny.
        d. White, disk x yellow, sphere gives 1/4 white, disk, 1/4 white, sphere, 1/4 yellow, disk, and 1/4 yellow, sphere.
        e. White, disk x white, sphere gives 3/8 white, disk, 3/8 white, sphere, 1/8 yellow, disk, and 1/8 yellow, sphere.

Answer: a. *WW Dd* x *ww dd*
           b. *Ww dd* x *Ww dd*
           c. *ww DD* x *WW dd*
           d. *Ww Dd* x *ww dd*
           e. *Ww Dd* x *Ww dd*

2.14   Genes *a*, *b*, and *c* assort independently and are recessive to their respective alleles *A*, *B*, and *C*. Two triply heterozygous (*Aa Bb Cc*) individuals are crossed.
        a. What is the probability that a given offspring will be phenotypically *ABC*, that is, exhibit all three dominant traits?
        b. What is the probability that a given offspring will be genotypically homozygous for all three dominant alleles?

Answer: The cross is *Aa Bb Cc* x *Aa Bb Cc*.
        a. Considering the *A* gene alone, the probability of an offspring showing the *A* trait from *Aa* x *Aa* is 3/4. Similarly the probability of showing the *B* trait from *Bb* x *Bb* is 3/4 and the *C* trait from *Cc* x *Cc* is 3/4. Therefore, the probability of a given progeny being phenotypically *ABC* (using the product rule) is 3/4 x 3/4 x 3/4 = 27/64.
        b. Considering the *A* gene, the probability of an *AA* offspring from *Aa* x *Aa* is 1/4. The same probability is the case for a *BB* offspring and for a *CC* offspring. Therefore, the probability of an *AA BB CC* offspring (from the product rule) is 1/4 x 1/4 x 1/4 = 1/64.

13

2.15   In garden peas tall stem *(T)* is dominant over short stem *(t)*, green pods *(G)* are dominant over yellow pods *(g)*, and smooth seeds *(S)* are dominant over wrinkled seeds *(s)*. Suppose a homozygous short, green, wrinkled pea plant is crossed with a homozygous tall, yellow, smooth one.
   a. What will be the appearance of the F_1?
   b. What will be the appearance of the F_2?
   c. What will be the appearance of the offspring of a cross of the F_1 back to its short, green, wrinkled parent?
   d. What will be the appearance of the offspring of a cross of the F_1 back to its tall, yellow, smooth parent?

Answer: The cross is *tt GG ss* x *TT gg SS*.
   a. The F_1 is triply heterozygous *Tt Gg Ss*, which will show the dominant phenotype of each gene pair and hence will have a tall stem, green pods, and smooth seeds.
   b. This cross is a trihybrid cross. The number of expected phenotypic classes is $2^3 = 8$. For each gene pair we expect a 3:1 ratio of F_2 progeny with dominant and recessive characteristics, respectively. Each gene pair segregates independently. Therefore the answer is 27/64 tall, green, smooth; 9/64 tall, green, wrinkled; 9/64 tall, yellow, smooth; 9/64 short, green, smooth; 3/64 tall, yellow, wrinkled; 3/64 short, green, wrinkled; 3/64 short, yellow, smooth; 1/64 short, yellow, wrinkled.
   c. The cross is *Tt Gg Ss* x *tt GG ss*. Half the progeny will be tall, and half will be short. All the progeny will be green. Half the progeny will be smooth, and half will be wrinkled. The compiled answer is 1/4 tall, green, smooth; 1/4 tall, green, wrinkled; 1/4 short, green, smooth; 1/4 short, green, wrinkled.
   d. The cross is *Tt Gg Ss* x *TT gg SS*. All the progeny will be tall and smooth. Half the progeny will be green, and half will be yellow.

2.16   Two homozygous strains of corn are hybridized. They are distinguished by six different pairs of genes, all of which assort independently and produce an independent phenotypic effect. The F_1 hybrid is selfed to give an F_2.
   a. What is the number of possible genotypes in the F_2?
   b. How many of these genotypes will be homozygous at all six gene loci?
   c. If all gene pairs act in a dominant-recessive fashion, what proportion of the F_2 will be homozygous for all dominants?
   d. What proportion of the F_2 will show all dominant phenotypes?

Solution: 1. The number of genotypes in any F_2 can be predicted by the expression 3^n, where n = number of genes (allelic pairs) in question. Hence, for a monohybrid cross there is one gene, so $n = 1$; therefore $3^n = 3$. For six gene loci, therefore, $n = 6$, and the answer to part "a" is 729.
   2. The offspring of any monohybrid cross should consist of one quarter homozygous for one allele and one quarter homozygous for the other allele. It follows then that two quarters, or one half, should consist of homozygotes (the remaining half being heterozygous). According to the law of coincident probability, the product rule applies, so that one would predict that $(1/2)^6$, where $n=6$ would represent the proportion of offspring that would be homozygous at all six gene loci; that is, 1/64.
   3. By similar reasoning, 1/4 of the offspring of a monohybrid cross involving a dominant allele would be predicted to be homozygous dominant. Therefore, again applying the product rule, the probability of being homozygous dominant for six gene pairs would be $(1/4)^6$, or 1/4096.
   4. If 3/4 is the proportion of the F_2 of a monohybrid cross that would show the dominant trait, what would be the proportion of the F_2 of a hexa(6)hybrid cross? Using the same logic as above, we get $(3/4)^6 = 729/4096$.

14

Answer: a. 729
              b. 1/64
              c. 1/4096
              d. 729/4096

2.17   The coat color of mice is controlled by several genes. The agouti pattern, characterized by a yellow band of pigment near the tip of the hairs, is produced by the dominant allele *A*; homozygous *aa* mice do not have the band and are nonagouti. The dominant allele *B* determines black hairs, and the recessive allele *b* determines brown. Homozygous $c^h c^h$ individuals allow pigments to be deposited only at the extremities (e.g., feet, nose, and ears) in a pattern called Himalayan. The genotype *C–* allows pigment to be distributed over the entire body.
        a.   If a true-breeding black mouse is crossed with a true-breeding brown, agouti, Himalayan mouse, what will be the phenotypes of the F_1 and F_2?
        b.   What proportion of the black agouti F_2 will be of genotype *Aa BB Cc^h*?
        c.   What proportion of the Himalayan mice in the F_2 are expected to show brown pigment?
        d.   What proportion of all agoutis in the F_2 are expected to show black pigment?

Answer: a.   The F_1 has the genotype *Aa Bb Cc^h* and is all agouti, black. The F_2 is 27/64 agouti, black; 9/64 agouti, black, Himalayan; 9/64 agouti, brown; 9/64 black; 3/64 agouti, brown, Himalayan; 3/64 black, Himalayan; 3/64 brown; 1/64 brown, Himalayan.
              b.   27/64 of the F_2 are agouti, black. From the F_1 x F_1 cross, *Aa Bb Cc^h* x *Aa Bb Cc^h* the proportion of F_2 mice with the genotype *Aa BB Cc^h* = 1/2 x 1/4 x 1/2 = 1/16 = 4/64. Therefore, the proportion of F_2 agouti, blacks that are *Aa BB Cc^h* = 4/64 divided by 27/64 = 4/64 x 64/27 = 4/27.
              c. F_2 Himalayans = 9/64 (agouti, black, Himalayan) + 3/64 (agouti, brown, Himalayan) + 3/64 (black, Himalayan) + 1/64 (brown, Himalayan) = 16/64 Himalayan. There are 3/64 + 1/64 = 4/64 brown, Himalayans; hence, the proportion of F_2 Himalayans that are brown = 4/64 x 64/16 = 4/16 = 1/4.
              d. F_2 agoutis = 27/64 (agouti, black) + 9/64 (agouti, black, Himalayan) + 9/64 (agouti, brown) + 3/64 (agouti, brown, Himalayan) = 48/64. There are 27/64 + 9/64 = 36/64 F_2 black, agoutis. Therefore, the proportion of F_2 agoutis that are black is 36/48 = 3/4.

2.18   In cocker spaniels, solid coat color is dominant over spotted coat. Suppose a true-breeding, solid-colored dog is crossed with a spotted dog, and the F_1 dogs are interbred.
        a.   What is the probability that the first puppy born will have a spotted coat?
        b.   What is the probability that if four puppies are born, all of them will have a solid coat?

Solution: 1.   True-breeding lines are assumed empirically to be homozygous, and an individual displaying a recessive trait must also be homozygous. Therefore the F_1 should be heterozygous, and interbreeding its members constitutes a monohybrid cross.
              2.   The puppies resulting from this monohybrid cross should assort out in the ratio of 75% solid to 25% spotted. Therefore the probability of a spotted puppy would be 25%, or 1/4, regardless of whether it was first, middle, or last in order of birth.
              3.   The product rule applies to solving part (b): namely, the product of 1/4 x 1/4 x 1/4 x 1/4, i.e. $(1/4)^4$ = 1/256, or 0.39 %.

Answer: a.   1/4
              b.   1/256

CHAPTER 3

CHROMOSOMAL BASIS OF INHERITANCE, SEX DETERMINATION, AND SEX LINKAGE

I. CHAPTER OUTLINE

Chromosome Theory of Inheritance
    Sex Chromosomes
    Sex Linkage
    Nondisjunction of X Chromosomes
Sex Determination and Sex Linkage in Eukaryotic Systems
    Genotypic Sex Determination Systems
    Phenotypic Sex Determination Systems
    Sex Linkage in Humans

II. IMPORTANT TERMS AND CONCEPTS

Chromosome Theory of Inheritance	meta- & telocentric
X chromosome	Testis-determining factor (TdF)
Y chromosome	Turner syndrome
heterogametic sex	Klinefelter syndrome
homogametic sex	Triplo-X syndrome
wild type	XYY-syndrome
hemizygous	dosage compensation
crisscross inheritance	antibody
sex-linked (X-linked)	antigen
autosome	isogenic
primary nondisjunction	Barr body
secondary nondisjunction	lyonization
aneuploidy	histocompatibility
genotypic sex determination system	H-Y antigen
phenotypic (environmental)	Z and W chromosomes
sex determination	monoecious *vs.* dioecious
X chromosome-autosome	hermaphrodite
balance system	mating type
chromosomal mechanism of	*MATa* & *MATα* alleles
sex determination	Y-linked (holandric) inheritance

III. THINKING ANALYTICALLY

The problems in this chapter are more complicated than the ones you encountered in Chapter 2, and so it is more important than ever to approach your solutions analytically and systematically. Always, after you have arrived at a solution, go back and test it against the data presented in the problem.

IV. QUESTIONS FOR PRACTICE

A. Multiple Choice Questions

1. The chromosome theory of inheritance holds that
    a. chromosomes are inherited.
    b. the chromosomes contain the hereditary material.
    c. the genes are inherited.
    d. the chromosomes are DNA.

2. Independent assortment of genes occurs if
    a. they are located on the sex chromosomes.
    b. they consist of different alleles.
    c. they reside in homologous chromosomes.
    d. they are located on different chromosome pairs.

3. Normally alleles are segregated into separate cells
    a. at the time of fertilization.
    b. during meiosis.
    c. by mitosis.
    d. by crossing over.

4. An individual that produces both ova and sperm
    a. has the XY genotype.
    b. reproduces asexually.
    c. is said to be hermaphroditic.
    d. both b and c, above.

5. The somatic cells of an individual having the karyotype 48, XXXY would have
    a. four Barr bodies.
    b   three Barr bodies.
    c. two Barr bodies.
    d. no Barr bodies.

Answers: 1b, 2d, 3b, 4c, 5c

B. Thought Questions

1. What is meant by crisscross inheritance, and of what is it diagnostic? (cf. text, p. 69)
2. What is the genetic basis of histocompatibility, and what clinical significance does it have? (cf. text, pp. 79-80.)
3. Define the term "lyonization", give an example of it, and discuss its functional significance. (cf. text, p.78, and Chapter 17)
4. Why do you think that most flowering plants are monoecious, whereas in most animals the sexes are separate? (Hint: Might motility be one consideration?)
5. If all mammals have evolved from a common ancestor, how do you account for the variety of chromosome numbers that they possess? By what mechanisms or mechanisms do you think changes in chromosome number could have occurred? (Hint: Think of chromosomal anomolies, such as aneuploidy, nondisjunction, and perhaps other accidental changes.)

V. ANSWERS AND SOLUTIONS TO TEXT QUESTIONS

3.1 At the time of synapsis preceding the reduction division in meiosis, the homologous chromosomes align in pairs and one member of each pair passes to each of the daughter nuclei (see Chapter 1). In an animal with five pairs of chromosomes, assume that chromosomes 1, 2, 3, 4, and 5 have come from the father, and 1', 2', 3', 4', and 5' have come from the mother. In what proportion of the germ cells of this animal will all the paternal chromosomes be present together?

Answer: The probability of a given homolog going to one particular pole is 1/2. The probability of all five paternal chromosomes going to the same pole is $(1/2)^5 = 1/32$. The same answer applies for all maternal chromosomes going to one pole.

3.2 In a male *Homo sapiens* from which grandparent could each sex-chromosome have been derived? (indicate 'Yes' or 'No' for each option.)

	Mother's		Father's	
	Mother	Father	Mother	Father
X chromosome	_____	_____	_____	_____
Y chromosome	_____	_____	_____	_____

Solution: A straightforward and logical way to solve this problem is to make a diagram of each generation using symbols X and Y, starting with the male in question and working up to his parents and then grandparents, thus:

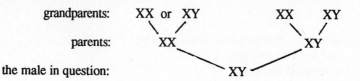

Answer:

	Mother's		Father's	
	Mother	Father	Mother	Father
X chromosome	Yes	Yes	No	No
Y chromosome	No	No	No	Yes

3.3 In *Drosophila*, white eyes are a sex-linked character. The mutant allele for white eyes (w) is recessive to the wild-type allele for brick red eye color (w^+).
   a. A white-eyed female is crossed with a red-eyed male. An F_1 female from this cross is mated with her father, and an F_1 male is mated with his mother. What will be the appearance of the offspring of these last two crosses with respect to eye color?
   b. A white-eyed female is crossed with a red-eyed male, and the F_2 from this cross is interbred. What will be the eye color of the F_3?

Answer:
   a. The original cross is $ww \times w^+Y$. The F_1 female is w^+w and is crossed with the w^+Y parental male. All female progeny of this cross will have red eyes; half the male progeny will have white eyes, and half will have red eyes. The F_1 male is wY and is crossed with the ww parental female. All progeny of this cross will have white eyes.
   b. The P_1 cross is $ww \times w^+Y$. The F_1 is w^+w female and wY male. Interbreeding the F_1 gives females, half having red eyes and half having white eyes. Half the males in the F_2 have red eyes and half have white eyes. Interbreeding the F_2 involves two types of females with two types of males. In the F_3, 5/16 of the progeny are red-eyed females, 3/16 are white-eyed females, 2/16 are red-eyed males, and 6/16 are white-eyed males.

3.4 One form of color blindness (c) in humans is caused by a sex-linked recessive mutant gene. A woman with normal color vision (c^+) and whose father was color-blind marries a man of normal vision whose father was also color-blind. What proportion of their offspring will be color-blind? (Give your answer separately for males and females.)

Answer: The woman is heterozygous for the recessive color blindness allele, let us say c^+c. The man is not color-blind and thus has the genotype c^+Y. (It does not matter what phenotype his mother and father have since males are hemizygous for sex-linked genes; thus the genotype can be assigned directly from the phenotype.) All the female offspring will have normal color vision; half the male offspring will be color-blind, and half will have normal color vision.

3.5   In humans, red-green color blindness is due to an X-linked recessive gene. A color-blind daughter is born to a woman with normal color vision and a father who is color-blind. What is the mother's genotype with respect to the alleles concerned?

Answer: The mother must be heterozygous for the red-green color blindness allele. If c is the symbol for red-green color blindness, her husband is cY and their daughter is cc. The mother has normal vision but transmits to her daughter an X chromosome with the mutant red-green color blindness allele. Therefore, the mother is a carrier for the trait; genotypically c^+c.

3.6   In humans, red-green color blindness is recessive and X-linked, while albinism is recessive and autosomal. What types of children can be produced as the result of marriages between two homozygous parents, a normal-visioned albino woman and a color-blind, normally pigmented man?

Answer: If c is the red-green color blindness allele, and a is the albino allele, the parent's genotypes are $c^+c^+\ aa$ ♀ and $cY\ a^+a^+$ ♂. All children will be normal visioned $(c^+c$ and $c^+Y)$ and normally pigmented $(a^+a$, both sexes).

3.7   In *Drosophila*, vestigial (partially formed) wings (vg) are recessive to normal long wings (vg^+), and the gene for this trait is autosomal. The gene for the white eye trait is on the X chromosome. Suppose a homozygous white, long-winged female fly is crossed with a homozygous red-eyed, vestigial-winged male.
   a.   What will be the appearance of the F_1?
   b.   What will be the appearance of the F_2?
   c.   What will be the appearance of the offspring of a cross of the F_1 back to each parent?

Answer: The parentals are $ww\ vg^+vg^+$ ♀ and $w^+Y\ vg\ vg$ ♂.
   a.   The F_1 males are all $wY\ vg^+vg$, white eyes, long wings. The F_1 females are all $w^+w\ vg^+vg$, red eyes, long wings.
   b.   The F_2 females are 3/8 red, long; 3/8 white, long, females; 1/8 red, vestigial, females; 1/8 white, vestigial, females; the same ratios of the respective phenotypes apply for the males.

3.8   In *Drosophila*, two red-eyed, long-winged flies are bred together and produce the following offspring: Females are 3/4 red, long and 1/4 red, vestigial; males are 3/8 red, long, 3/8 white, long, 1/8 red, vestigial, and 1/8 white, vestigial. What are the genotypes of the parents?

Solution: 1.   A red-eyed, long-winged female must have at least one w^+ allele and one vg^+ allele, so tentatively her genotype is $w^+-\ vg^+-$. The male, similarly, must be $w^+Y\ vg^+-$.
   2.   If the female offspring assort 3 long to 1 vestigial, both parents must be heterozygous for the vg gene, and therefore are $w^+-\ vg^+vg$ and $w^+Y\ vg^+vg$.
   3.   Notice that some of the males are white-eyed. This allele must have come from the female parent, and so she must be heterozygous at the white locus.

Answer: The female genotype is $w^+w\ vg^+vg$, and the male genotype is $w^+Y\ vg^+vg$.

3.9 In poultry a dominant sex-linked gene (*B*) produces barred feathers, and the recessive allele (*b*), when homozygous, produces nonbarred feathers. Suppose a nonbarred cock is crossed with a barred hen.
    a.   What will be the appearance of the F_1 birds?
    b.   If an F_1 female is mated with her father, what will be the appearance of the offspring?
    c.   If an F_1 male is mated with his mother, what will be the appearance of the offspring?

Answer: In chickens the homogametic sex is the male, and the heterogametic sex is the female. The original cross therefore is BW ♀ x *bb* ♂ (the W is the unpaired sex chromosome).
    a.   The F_1 birds are barred males *(Bb)* and nonbarred females *(bW)*.
    b.   The cross is *b*W x *bb*, to give all nonbarred progeny.
    c.   The cross is *Bb* x *B*W, to give all males barred, half the females barred, and half the females nonbarred.

3.10 A man (A) suffering from defective tooth enamel, which results in brown-colored teeth, marries a normal woman. All their daughters have brown teeth, but the sons are normal. The sons of man A marry normal women, and all their children are normal. The daughters of man A marry normal men, and 50 percent of their children have brown teeth. Explain these facts.

Answer: The simplest hypothesis is that brown-colored teeth are determined by a sex-linked dominant mutant allele. Man A was *B*Y and his wife was *bb*. All sons will be *b*Y normals and cannot pass on the trait. All the daughters receive the X chromosome from their father, so they are *Bb* with brown enamel. Half their sons will have brown teeth since half their sons receive the X chromosome with the *B* mutant allele.

3.11 In humans, differences in the ability to taste phenylthiourea are due to a pair of autosomal alleles. Inability to taste is recessive to ability to taste. A child who is a nontaster is born to a couple who can both taste the substance. What is the probability that their next child will be a taster?

Answer: To produce a recessive, nontaster child while both being tasters must mean that both parents are heterozygous for the trait; that is, t^+t if t is the nontaster allele. From pairings of two heterozygotes, 3/4 of the progeny should exhibit the dominant phenotype and 1/4 the recessive phenotype. The answer is 3/4.

3.12 Cystic fibrosis is inherited as an autosomal recessive. Two noncystic fibrosis parents have two children with cystic fibrosis and three children who do not have cystic fibrosis. They come to you for genetic counseling.
    a.   What is the numerical probability that their next child will have cystic fibrosis?
    b.   Their non-affected children are concerned about being heterozygous. What is the numerical probability that a given non-affected child in the family is heterozygous?

Answer: a.   To have produced a child with cystic fibrosis, both parents must have been heterozygous for the mutant gene. Therefore, the probability of their next child having cystic fibrosis is 1/4, or one in four.
    b.   Non-affected children can be either *AA* or *Aa*. From a cross *Aa* x *Aa* there will be a progeny ratio of 1 *AA*:2 *Aa*:1 *aa*. Therefore, the probability that a non-affected child will be heterozygous is 2/3.

3.13 Huntington's disease is a human disease inherited as a Mendelian autosomal dominant. The disease results in choreic (uncontrolled) movements, progressive mental deterioration, and eventually death. The disease affects the carriers of the trait anytime between 15 and 65 years of age. The American folksinger Woody Guthrie died of Huntington's disease, as did one of his parents. Marjorie Mazia, Woody's wife, had no history of this disease in her family. The Guthries had three children. What is the probability that a particular Guthrie child will die of Huntington's disease?

Answer: Since only one of Woody's parents died of this disease, Woody must have been heterozygous *Cc* for the dominant mutant allele. His wife is *cc*. From *Cc* x *cc* the probability of having a child that will die of Huntington's chorea is 1/2.

3.14 Suppose gene A is on the X chromosome, and genes B, C, and D are on three different autosomes. Thus $A-$ signifies the dominant phenotype in the male or female. An equivalent situation holds for $B-$, $C-$, and $D-$. The cross $AA\ BB\ CC\ DD$ females x $aY\ bb\ cc\ dd$ males is made.

    a. What is the probability of obtaining an $A-$ individual in the F_1?

    b. What is the probability of obtaining an a male in the F_1?

    c. What is the probability of obtaining an $A-B-C-D-$ female in the F_1?

    d. How many different F_1 genotypes will there be?

    e. What proportion of F_2's will be heterozygous for the four genes?

    f. Determine the probabilities of obtaining each of the following types in the F_2: (1) $A-bb$ $CC\ dd$ (female); (2) $aY\ BB\ Cc\ Dd$ (male); (3) $AY\ bb\ CC\ dd$ (male); (4) $aa\ bb\ Cc\ Dd$ (female).

Answer: a. The cross is AA x aY, so all progeny are $A-$;, thus the probability of $A-$ is 1.

    b. 0

    c. All females will have the phenotype required.

    d. There are three genotypes for each of the autosomal genes, and four genotypes for the sex-linked gene (AA, Aa, AY, aY), giving $4(3)^3 = 108$.

    e. Only females can be heterozygous for the sex-linked gene, and the probability of a female offspring is 1/2. The probability of heterozygosity for any particular gene pair is 1/2, and therefore for four gene pairs the answer is $(1/2)^4$ x 1/2 (for the probability of femaleness) = 1/32.

    f. (1) 1/64 of females (1/128 of total progeny)

        (2) 1/32 of males (1/64 of total progeny)

        (3) 1/128 of males (1/256 of total progeny)

        (4) 0

3.15 As a famous mad scientist, you have cleverly devised a method to isolate *Drosophila* ova that have undergone primary nondisjunction of the sex chromosomes. In one experiment you used females homozygous for the sex-linked recessive mutation causing white eyes (*w*) as your source of nondisjunction ova. The ova were collected and fertilized with sperm from red-eyed males. The progeny of this "engineered" cross were then backcrossed separately to the two parental strains (this is called "backcrossing"). What classes of progeny (genotype and phenotype) would you expect to result from these backcrosses? (The genotype of the original parents may be denoted as ww for the females and w^+Y for the males.)

Answer: The answer is given in the following Punnett squares:

ww x w^+Y

Sperm

Eggs	w^+	Y
ww	www^+	wwY
0	w^+0	$Y0$

Survivors are wwY (white ♀) and w^+0 (sterile, red ♂)

Backcross wwY x w^+Y

Eggs		Sperm: w^+	Sperm: Y
Normal	w Y	$w^+ wY$ Red ♀	w YY White ♂
Normal	w	$w^+ w$ Red ♀	w Y White ♂
Secondary nondisjunction	ww	$w^+ ww$ Triplo-X Usually dies	ww Y White ♀
Secondary nondisjunction	Y	w^+ Y Red ♂	Y Y Dies

3.16 In *Drosophila* the bobbed gene (bb^+) is located on the X chromosome. Unlike most X-linked genes, however, the Y chromosome also carries a bobbed gene. The mutant allele *bb* is recessive to bb^+. If a wild-type F_1 female that resulted from primary nondisjunction in oogenesis in a cross of bobbed female with a wild-type male is mated to a bobbed male, what will be the phenotypes and their frequencies in the offspring? List males and females separately in your answer. (Hint: Refer to the text for information about the frequency of nondisjunction in *Drosophila*.)

Answer: The answer is given in the following figure:

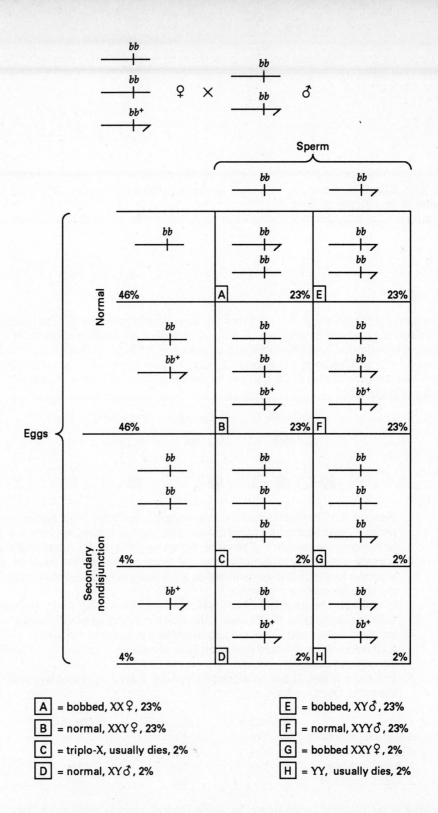

A = bobbed, XX ♀, 23%

B = normal, XXY ♀, 23%

C = triplo-X, usually dies, 2%

D = normal, XY ♂, 2%

E = bobbed, XY ♂, 23%

F = normal, XYY ♂, 23%

G = bobbed XXY ♀, 2%

H = YY, usually dies, 2%

In sum, 4% lethal: 25% bobbed females; 23% normal females; 23% bobbed males; 25% normal males.

3.17 A Turner syndrome individual would be expected to have the following number of Barr bodies in the majority of cells:
   a. 0
   b. 1
   c. 2
   d. 3

Answer: a

3.18 An XXY Klinefelter syndrome individual would be expected to have the following number of Barr bodies in the majority of cells:
   a. 0
   b. 1
   c. 2
   d. 3

Answer: b

3.19 In human genetics the pedigree is used for analysis of inheritance patterns. The female is represented by a circle and the male by a square. The figure presents three family pedigrees for a trait in human beings. Normal individuals are represented by unshaded symbols, and people with the trait by shaded symbols. For each pedigree (A, B, and C), state, by answering yes or no in the appropriate blank space whether transmission of the trait can be accounted for on the basis of each of the listed simple modes of inheritance:

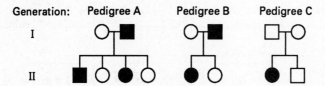

Solution:
   1. Pedigree A: The mother could be heterozygous, the father homozygous recessive, and the children could be a homozygous recessive boy, two heterozygous girls, and one girl would be homozygous recessive. Therefore, the answer is "Yes" for Autosomal Recessive.
   2. It could also be an autosomal dominant, because the parents could be a homozygous recessive mother, a heterozygous father, a boy and girl heterozygous (symbols shaded), and two girls homozygous recessive.
   3. It could also be an X-linked recessive, because the mother could be heterozygous, the father recessive (plus Y, of course), the son recessive, two of the daughters heterozygous, and the daughter who shows the trait would be homozygous recessive.
   4. It could not be an X-linked dominant, however, because the trait could not be passed from the father to a son.
   5. Pedigrees B and C can be solved by similar analytical reasoning, with the following answer to Question 3.19.

Answer:

	Pedigree A	Pedigree B	Pedigree C
Autosomal recessive	Yes	Yes	Yes
Autosomal dominant	Yes	Yes	No
X-linked recessive	Yes	Yes	No
X-linked dominant	No	No	No

3.20 Looking at the pedigree in the figure, in which shaded symbols represent a "trait," which of the progeny (as designated by numbers) eliminate X-linked recessiveness as a mode of inheritance for the trait?
   a. 1 and 2
   b. 4
   c. 5
   d. 2 and 4

Generation:

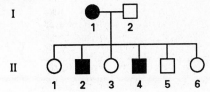

I

II

Answer: c is the correct answer. If the mother has an X-linked recessive trait, then all sons must have the trait.

3.21 When constructing human pedigrees, geneticists often refer to particular persons by a number. The generations are labeled by roman numerals and the individuals in each generation by arabic numerals. For example, in the pedigree in the figure, the female with the asterisk would be I.2. Use this means to designate specific individuals in the pedigree. Determine the probable mode for the trait shown in the affected individuals (the shaded symbols) by answering the following questions. Assume the condition is caused by a single gene.

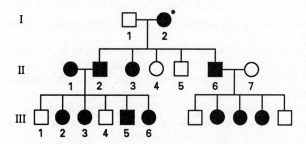

a. Y-linked inheritance can be excluded at a glance. What two other mechanisms of inheritance can be definitely excluded? Why can these be excluded?
b. Of the remaining mechanisms of inheritance, which is the most likely? Why?

Answer:
a. X-linked recessive can be excluded, for example, because individual I.2 would be expected to pass on the trait to all sons and that is not the case. Also, the II.1, II.2 pairing would be expected to produce offspring all of whom express the trait. Autosomal recessive can also be excluded because the II.1, II.2 pairing would be expected to produce offspring all of whom express the trait.
b. The remaining two mechanisms of inheritance are X-linked dominant and autosomal dominant. Genotypes can be written to satisfy both mechanisms of inheritance. X-linked dominance is perhaps more likely since the II.6, II.7 pairing shows exclusively father-daughter inheritance, that is, all of the daughters and none of the sons exhibit the trait: this is a characteristic of the segregation of X-linked dominant alleles. If the trait were autosomal dominant, then half of the sons and half of the daughters would be expected to exhibit the trait.

3.22   Phenylketonuria (PKU) is an inborn error in the metabolism of the amino acid phenylalanine. The
characteristic feature of PKU is severe mental retardation; many untreated patients with this trait
have IQs below 20. The three-generation pedigree shown in the figure is of an affected family.

Generation:

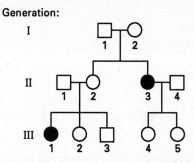

a.   What is the mechanism of inheritance of PKU?
b.   Which persons in the pedigree are known to be heterozygous for PKU?
c.   What is the probability that III.2 is a carrier (heterozygous)?
d.   If III.3 and III.4 marry, what is the probability that their first child will have PKU?

Solution: 1. Since the trait skips generations, it must be due to a autosomal recessive gene.
2. Since it appears only in the females, it must be autosomal. Therefore the answer to part
1a is "autosomal recessive".
3. Both I.1 and I.2 must be carriers for the allele for it to be expressed in II.3. II.1 and II.2
must also be carriers for it in order for it to be expressed in III.1. III.4 and III.5 must be
carriers too because II.3 is homozygous. Since carriers are necessarily heterozygous, the
answer to part 'b' is "I.1, I.2, II.1, II.2, III.4 and III.5".
4. The individuals designated II.1 and II.2 must both be heterozygous, therefore their
offspring would assort ideally in the ratio 1 homozygous dominant:2 heterozygous:1
homozygous recessive. Since the heterozygotes constitute 2/3 of the normals, and III.2 is
normal, the probability that she is heterozygous is 2/3, which is the answer to part 'c'.
5. By the same reasoning, III.3 has a 2/3 probability of being heterozygote. III.4 must be
heterozygote, because her mother was homozygous recessive. Since the probability that
two heterozygotes will have a homozygous recessive offspring is 1/4, and since III.3 has a
probability of 3/4 for heterozygosis, the product rule predicts a probability of 1/4 x 2/3 =
2/12 = 1/6 that any child this couple might have will have PKU. Hence the answer to
part 'd' is "1/6".

Answer: a. autosomal recessive
b. I.1,2
c. 2/3
d. 1/6

3.23 For the more complex pedigrees shown in the figure, indicate the probable mode of inheritance: autosomal recessive, autosomal dominant, X-linked recessive, Y-linked dominant, Y-linked.

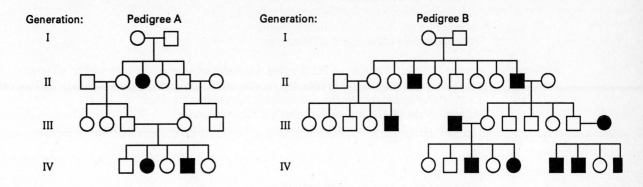

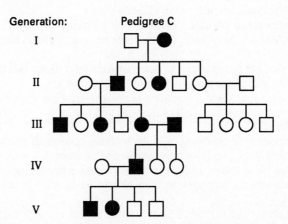

Solution: Pedigree A: At first glance this pedigree looks like an autosomal recessive. Certainly it is a recessive, because the trait skips generations; if it were dominant, it would either be evident in each generation or it would disappear altogether. It must be autosomal, because if it were X-linked it would have to be either in the father, which it is not (he would show it, if it were) or in the mother. If it were in the mother, it would have to be homozygous in the daughter (II.3), which is not possible if it was not in the father (I.2). This conclusion is supported by the absence of the trait in generation III, and reappearance in generation IV, in both sexes. *Answer:* autosomal recessive.

Pedigree B: Again we appear to be dealing with a recessive gene, because it was not evident in either parent (Generation I) . Of course it may have arisen *de novo* as a dominant mutation in the germ cells of one of these parents. The probability of this is considerably less than the probability that it is a recessive gene.

It also looks like an X-linked recessive, because it appears predominantly in males, and when it does appear in females (III), it is in an unrelated one or, as in IV.5, it is easily explained. The clincher is the appearance of the trait in all three sons (IV.6, IV.7, IV.9) of the homozygous female in III.12. *Answer:* X-linked recessive.

Pedigree C: This gene is expressed in every generation, and equally by males and females. It is clearly an autosomal dominant. If it were recessive, it would appear in IV.3 and IV.4; and if it were X-linked, it would also appear in these individuals. *Answer:* autosomal dominant.

Answer: Pedigree A: autosomal recessive; Pedigree B: X-linked recessive; Pedigree C: autosomal dominant.

3.24   If a rare genetic disease is inherited on the basis of an X-linked dominant gene, one would expect to find the following:
   a.   Affected fathers have 100% affected sons.
   b.   Affected mothers have 100% affected daughters.
   c.   Affected fathers have 100% affected daughters.
   d.   Affected mothers have 100% affected sons.

Answer: a is false: the father passes on his X to his daughters so *none* of the sons will be affected.
   b is false: since the disease is rare, the mother would be heterozygous and only 50% of her daughters would receive the X with the dominant mutant allele.
   c is true: all daughters receive the father's X chromosome.
   d is false: only 50% of her sons would receive the X with the dominant mutant allele.

3.25   If a genetic disease is inherited on the basis of an autosomal dominant gene, one would expect to find the following:
   a.   Affected fathers have only affected children.
   b.   Affected mothers never have affected sons.
   c.   If both parents are affected, all of their offspring have the disease.
   d.   If a child has the disease, one of his or her grandparents also had the disease.

Answer: a is false: the father could be heterozygous (and likely will be if the disease is rare) and hence only half of the progeny should have the disease.
   b is false: autosomal inheritance is regardless of the sex of the individual. If the mother was heterozygous, then half of the sons and half of the daughters would be expected to have the disease.
   c is false: both parents could be heterozygous (and likely will be if the disease is rare). Pairings of two heterozygotes would produce 3/4 with the dominant phenotype - in this case the disease - and 1/4 with the recessive phenotype, that is, normal individuals.
   d is true: for a dominant trait there should be no skips in the generations.

3.26   If a genetic disease is inherited as an autosomal recessive, one would expect to find the following:
   a.   Two affected individuals never have an unaffected child.
   b.   Two affected individuals have affected male offspring but no affected female children.
   c.   If a child has the disease, one of his or her grandparents will have had it.
   d.   In a marriage between an affected individual and an unaffected one, all the children are unaffected.

Answer: a is true, that is, *aa* x *aa* only produces *aa* progeny.
   b is false - since the disease is autosomal, the sex of the progeny makes no difference in the fact that all progeny will be affected if both parents are affected.
   c is false - a child must receive one autosomal recessive allele from each parent to have the disease but neither the parents nor the grandparents need have had the disease; all could theoretically be carriers.
   d is false: if the unaffected parent is a carrier, then affected progeny can result.

3.27   Which of the following statements is *not* true for a disease that is inherited as a rare X-linked dominant?
   a.   All daughters of an affected male will inherit the disease.
   b.   Sons will inherit the disease only if their mothers have the disease.
   c.   Both affected males and affected females will pass the trait to half their children.
   d.   Daughters will inherit the disease only if their father has the disease.

Answer: d is not true: a daughter could inherit that disease from her mother if the mother had the trait (genotype = *Aa).*

3.28 Women who were known to be carriers of the X-linked, recessive, hemophilia gene were studied in order to determine the amount of time required for the blood clotting reaction. It was found that the time required for clotting was extremely variable from individual to individual. The values obtained ranged from normal clotting time at one extreme all the way to clinical hemophilia at the other extreme. What is the most probable explanation for these findings?

Answer: Since hemophilia is an X-linked trait, the most likely explanation is that random inactivation of X chromosomes (lyonization, see p. 78), produces individuals with different proportions of cells with the normal allele. Thus, if some women had only 40% of their cells with an active h^+ allele, and 60% with the h (hemophilia) allele, while other women had 60% of their cells with an active h^+ allele, and 40% with the h allele, there would be a significant difference in the amount of clotting factor these two individuals would make.

3.29 In a species of amphibian, sex can be reversed by placing hormones in the water where fertilized eggs are developing. Eggs exposed to estrogens develop as phenotypic females, whatever their genetic sex. while eggs exposed to androgens all develop as phenotypic mates.

In one experiment several phenotypic females raised in estrogen-containing water were mated to normal males. About half of these females produced only mate offspring, while the other half produced males and females in the expected 1:1 ratio.

In another experiment several phenotypic males raised in androgen-containing water were mated to normal females. About half of these mates produced the normal 1:1 ratio of daughters and sons, but the other half produced about three females for every mate.

    a.   How is sex determined in these amphibians? Explain your reasoning.

    b.   If the offspring from the mating above that yielded a ratio of 3 females to 1 male were crossed to normal mates, how many kinds of matings would there be? What proportion of the two sexes would be seen in the offspring of each kind of mating?

Answer: a.   Females are XY, and males are XX. (The data also fit the model that males are homozygous for a recessive sex-determining allele while females are heterozygous for that and for a dominant, female-determining allele.). In the mating of estrogen-raised female to normal males, half of the matings were the usual XY females x XX males, and gave the expected 50% XY and 50% XY offspring. The other half of matings were XX sex reversed (female) x normal XX (male) and could produce only XX (male) offspring. In the matings of androgen-raised males to normal females, half of the matings were the usual XX x XY, but the other half were XY sex-reversed (male) x XY female. These yielded 25% XX (male), 50% XY (female) and 25% YY (female).

    b.   There would be two kinds. 75% of the matings would be XX x XY and yield 1 female:1 male. 25% of the matings would be XX x YY and would yield all XY (female) offspring.

CHAPTER 4

EXTENSIONS OF MENDELIAN GENETIC ANALYSIS

I. CHAPTER OUTLINE

Multiple Alleles
    <u>Drosophila</u> Eye Color
    ABO Blood Group
Modifications of Dominance Relationships
    Incomplete Dominance
    Codomainance
Gene Interactions and Modified Mendelian Ratios
    Gene Interactions That Produce New Phenotypes
    Epistasis
Essential Genes and Lethal Alleles
The Environment and Gene Expression
    Effects of the Internal Environment
    Effects of the External Environment
    Twin Studies

II. IMPORTANT TERMS AND CONCEPTS

multiple alleles	lethal allele
wild-type allele	essential gene
complete dominance	penetrance
complete recessiveness	expressivity
incomplete (partial) dominance	phenocopy
codominance	dizygotic *vs.* monozygotic twins
epistasis	concordance *vs.* discordance
hypostasis	

III. THINKING ANALYTICALLY

The methods of reasoning used in this chapter are not substantively different from those in Chapter 2, although ratios are modified because of a spectrum of interactions at one or more loci. It is important to understand that the action of a gene or the interaction between different genetic loci can be affected in several ways simultaneously, and can therefore be variously labelled. Moreover, alternate phenotypes that are initially studied clinically may exhibit other, or more subtle, relationships when investigated biochemically, so that once again, a variety of labels may be applied to indicate the kind and degree of interaction that may occur within one set of alleles, between several sets of alleles, and between genetic and environmental influences. Finally, it is well to remember that a trait may be the final expression of the particular action of a set of alleles at one or more than one locus, but these genes, in turn, work only within the general environment that has been previously determined genetically and environmentally during the development of the organism.

IV. QUESTIONS FOR PRACTICE

A. Multiple Choice Questions

Using the list below, choose the kind of inheritance that best explains the data. Explain your choices.

    a.   recessive lethal allele
    b.   dominant lethal allele
    c.   incomplete dominance; two allelic pairs
    d.   duplicate recessive epistasis; two allelic pairs
    e.   dominant epistasis; two allelic pairs

1. When crossed, two pure-breeding varieties of white kernel corn yield progeny with purple kernels. When these purple progeny are selfed, a count of the kernels in the F_2 yields an average of 56 purple kernels; 44 white kernels per ear.

2. When pure-breeding wheat with red kernels is crossed to the pure-breeding white variety, the progeny has pink kernels. The 100 F_1 plants that further result, exhibit the following phenotypes: 6 red, 6 white, 25 light red, 25 light pink, 38 pink.

3. Yellow mice crossed to pure-breeding agouti mice produced a 50:50 ratio of yellow to agouti offspring. Crosses between yellow mice never produced 100% yellow progeny, but rather yielded 36 yellow and 15 agouti offspring.

Answers: 1d, 2c, 3b

B. Thought Questions

1. Distinguish between dominance and epistasis, and between epistasis and multiple alleles.
2. How would you distinguish between a gene with pleiotropic effects and the effects resulting from several sets of alleles? (Hint: Review independent assortment.)
3. In Tay-Sachs disease, heterozygotes, who are phenotypically normal, exhibit a level of the enzyme, hexoseaminidase A, that is midway between that contained in recessive homozygotes, who have no detectable enzyme, and the level in normal, dominant homozygotes. Explain why Tay-Sachs disease could be variously described as an example of incomplete dominance, co-dominance, and/or a recessive lethal.
4. In a very large natural population, could you ever assume that you know all of the allelic variants in a multiple allelic series? (Hint: Could new variants appear at any time?)
5. Devise a situation in which a multiple allelic series shows elements of co-dominance as well as lethality.
6. Could a lethal allele, either dominant or recessive, be incompletely penetrant? (Hint: Think about small variations in the timing of events during development.)
7. The following <u>concordance percentages</u> are well known:

Trait	Monozygotic twins	Dizygotic twins
eye color	99.6	28
manic depression	77	19
handedness (either)	79	77
measles	95	87

Comment on the relative importance of penetrance and gene-*vs.*-environmental interaction for each of these traits. (Hint: First compare concordances to predict degree of specific genetic involvement, then consider penetrance. Do not neglect general developmental background.)

V. ANSWERS AND SOLUTIONS TO TEXT QUESTIONS

4.1 In rabbits, C = agouti coat color, c^{ch} = chinchilla, C^h = Himalayan, and c = albino. The four alleles constitute a multiple allelic series. The agouti C is dominant to the three other alleles, c is recessive to all three other alleles, and chinchilla is dominant to Himalayan. Determine the phenotypes of progeny from the following crosses:
   a. $C/C \times c/c$
   b. $C/C^{ch} \times C/c$
   c. $C/c \times C/c$
   d. $C/c^h \times c^h/c$
   e. $C/c^h \times c/c$
   f. $c^{ch}/c^h \times c^h/c$
   g. $c^h/c \times c/c$
   h. $C/c^h \times c/c$
   i. $C/c^h \times C/c^{ch}$

Answer: a. all agouti
         b. 3/4 agouti, 1/4 chinchilla
         c. 3/4 agouti, 1/4 albino
         d. 1/2 agouti, 1/2 Himalayan
         e. 1/2 agouti, 1/2 Himalayan
         f. 1/2 chinchilla, 1/2 Himalayan
         g. 1/2 Himalayan, 1/2 albino
         h. 3/4 agouti, 1/4 Himalayan
         i. 3/4 agouti, 1/4 chinchilla

4.2 If a given population of diploid organisms contains three, and only three, alleles of a particular gene (say w, $w1$, and $w2$), how many different diploid genotypes are possible in the populations? List all possible genotypes of diploids (considering *only* these three alleles).

Answer: 6 possible genotypes: ww, $ww1$ $ww2$, $w1w1$, $w1w2$, $w2w2$

4.3 The genetic basis of the ABO blood types seems most likely to be:
   a. multiple alleles
   b. polyexpressive hemizygotes
   c. allelically excluded alternates
   d. three independently assorting genes

Answer: a

4.4 In humans the three alleles I^A, I^B, and I^O constitute a multiple allelic series that determine the ABO blood group system, as we described in this chapter. For the following problems, state whether the child mentioned can actually be produced from the marriage. Explain your answer.
   a. An O child from the marriage of two A individuals.
   b. An O child from the marriage of an A to a B.
   c. An AB child from the marriage of an A to an O.
   d. An O child from the marriage of AB to an A.
   e. An A child from the marriage of an AB to a B.

Answer: a. Yes; both A's could be I^A/I^O heterozygotes so that an I^O/I^O child could result.
         b. If A is I^A/I^O and B is I^B/I^O, then an O child can result
         c. One parent must donate an I^A, allele and the other an I^B allele for a child to be AB. The O parent is I^O/I^O so it is not possible to produce an O child.
         d. The parents are $I^A/I^B \times I^A/_-$. An O child cannot result from this marriage.
         e. Yes; if the B parent is I^B/I^O, then the A child would be I^A/I^O.

4.5　A man is blood type O,M. A woman is blood type A,M and her child is type A,MN. The aforesaid man cannot be the father of the child because:
   a.　O men cannot have type A children.
   b.　O men cannot have MN children.
   c.　An O man and an A woman cannot have an A child.
   d.　An M man and an M woman cannot have an MN child.

Answer:　　d. The parents' genotypes are: man, I^O/I^O M/M woman I^A/I^A(or I^A/I^O)M/M. While a type A child can result from such a pairing, it is impossible to produce an MN child.

4.6　A woman of blood group AB marries a man of blood group A whose father was group O. What is the probability that
   a.　their two children will both be group A?
   b.　one child will be group B and the other group O?
   c.　the first child will be a son of group AB and their second child a son of group B?

Answer:　　The woman's genotype is I^A/I^B and the man's genotype is I^A/I^O.
   a.　1/2 x 1/2 = 1/4.
   b.　Zero. A blood group O baby is not possible.
   c.　1/2 (probability of male) x 1/4 (probability of AB) x 1/2 (probability of male) x 1/4 (probability of B) = 1/64 probability that all four conditions will be fulfilled.

4.7　If a mother and her child belong to blood group O, what blood group could the father *not* belong to?

Answer:　　Blood group O is genotype I^O/I^O so father could not be AB.

4.8　A man of what blood group could not be a father to a child of blood type AB?

Answer:　　Blood type O, since genotype I^O/I^O.

4.9　In snapdragons, red flower color (R) is incompletely dominant to white (r); the R/r heterozygotes are pink. A red-flowered snapdragon is crossed with a white-flowered one. Determine the flower color of (a) the F_1; (b) the F_2; (c) the progeny of a cross of the F_1 to the red parent; (d) the progeny of a cross of the F_1 to the white parent.

Answer:　　a.　pink
   b.　1/4 red, 1/2 pink, 1/4 white
   c.　1/2 red, 1/2 pink
   d.　1/2 pink, 1/2 white

4.10　In Shorthorn cattle the heterozygous condition of the alleles for red coat color (R) and white coat color (r) is roan coat color. If two roan cattle are mated, what proportion of the progeny will resemble their parents in coat color?

Answer:　　Half will be R/r and hence will resemble the parents.

4.11   What progeny will a roan Shorthorn have if bred to (a) red; (b) roan; (c) white?

Solution: In Shorthorn cattle, the color roan indicates a heterozygote (R/r) because this pair of alleles exhibits incomplete dominance. Hence, the three crosses desired are:

a. R/r (roan) x R/R (red). Since 1/2 of the gametes of the roan parent will carry the R allele and 1/2 will contain the r allele, when fertilized by gametes, all of which contain the R allele, 1/2 of the offspring will be roan (R/r) and 1/2 will be red (R/R).

b. Roan x Roan is a typical monohybrid cross. It will yield a 1 R/R:2 R/r:1 r/r genotypic ration. Because of incomplete dominance this will be reflected in a phenotypic ratio that is 1 red:2 roan:1 white.

c. When a heterozygote is bred to a recessive homozygote, we have a cross that recalls one of the possibilities in a testcross. The R containing gametes from the roan parent, as well as the r containing gametes will all be fertilized by an r containing gamete. Hence, 1/2 of the offspring will be roan and 1/2 will be white.

Answer: a. 1/2 red, 1/2 roan

b. 1/4 red, 1/2 roan, 1/4 white

c. 1/2 roan, 1/2 white

4.12   In peaches, fuzzy skin (F) is completely dominant to smooth (nectarine) skin (f), and the heterozygous conditions of oval glands at the base of the leaves (O) and no glands (o) give round glands. A homozygous fuzzy, no-gland peach variety is bred to a smooth, oval-gland variety.

a. What will be the appearance of the F_1?

b. What will be the appearance of the F_2?

c. What will be the appearance of the offspring of a cross of the F_1 back to the smooth, oval-glanded parent?

Answer: a. The cross is $F/F\ o/o$ x $f/f\ O/O$. The F_1 is $F/f\ O/o$, which is fuzzy with round leaf glands.

b. Interbreeding the F_1 gives 3/16 fuzzy, oval-glanded; 6/16 fuzzy, round-glanded; 3/16 fuzzy, no-glanded; 1/16 smooth, oval-glanded; 2/16 smooth, round-glanded; 1/16 smooth, no-glanded.

c. Cross is $F/f\ O/o$ x $f/f\ O/O$. The progeny are 1/4 fuzzy, oval-glanded; 1/4 fuzzy, round-glanded; 1/4 smooth, oval-glanded; 1/4 smooth, round-glanded.

4.13   In guinea pigs, short hair (L) is dominant to long hair (l), and the heterozygous conditions of yellow coat (W) and white coat (w) give cream coat. A short-haired, cream guinea pig is bred to a long-haired, white guinea pig, and a long-haired, cream baby guinea pig is produced. When the baby grows up, it is bred back to the short-haired, cream parent. What phenotypic classes and in what proportions are expected among the offspring?

Answer: The parents were $L/l\ W/w$ and $l/l\ w/w$. The baby was $l/l\ W/w$. From a cross of $l/l\ W/w$ x $L/l\ W/w$ the phenotypic classes expected are 1/8 short, yellow; 2/8 short, cream; 1/8 short, white; 1/8 long, yellow; 2/8 long, cream; 1/8 long, white.

4.14   The shape of radishes may be long (L/L), oval (L/l), or round (l/l), and the color of radishes may be red (R/R), purple (R/r), or white (r/r). If a long, red radish plant is crossed with a round, white plant what will be the appearance of the F_1 and the F_2?

Answer: The cross is $L/L\ R/R$ x $l/l\ r/r$. The F_1 is $L/l\ R/r$, oval and purple. Selfing the doubly heterozygous F_1 gives the following phenotypic classes in the F_2: 1/16 long, red; 2/16 long, purple; 1/16 long, white; 2/16 oval, red; 4/16 oval, purple; 2/16 oval, white; 1/16 round, red; 2/16 round, purple; 1/16 round, white.

4.15   In poultry the genes for rose comb (*R*) and pea comb (*P*), if present together, give walnut comb. The recessive alleles of each gene, when present together in a homozygous state, give single comb. What will be the comb characters of the offspring of the following crosses?

a.   *R/R P/p* x *r/r P/p*
b.   *r/r P/P* x *R/r P/p*
c.   *R/r p/p* x *r/r P/p*
d.   *R/r P/p* x *R/r P/p*
e.   *R/r p/p* x *R/r p/p*

Answer: a.   3/4 walnut *(R/– P/–)* and 1/4 rose *(R/– p/p)*
              b.   1/2 walnut and 1/2 pea *(r/r P/–)*
              c.   1/4 walnut, 1/4 pea, 1/4 rose, 1/4 single
              d.   9/16 walnut, 3/16 rose, 3/16 pea, 1/16 single
              e.   3/4 rose, 1/4 single

4.16   For the following crosses involving the comb character in poultry, determine the genotypes of the two parents:

a.   A walnut crossed with a single produces offspring 1/4 of which are walnut, 1/4 rose, 1/4 pea, and 1/4 single.
b.   A rose crossed with a walnut produces offspring 3/8 of which are walnut, 3/8 rose, 1/8 pea, and 1/8 single.
c.   A rose crossed with a pea produces five walnut and six rose offspring.
d.   A walnut crossed with a walnut produces one rose, two walnut, and one single offspring.

Solution: As stated in the previous problem, comb character involves two sets of alleles, *R* (rose comb) and *P* (pea comb). Any genotype with at least one dominant from both pairs (*R/– P/–*) produces a walnut comb. The double recessive (*r/r pp*) results in a single comb. From chapter data, plus inference, assume that *R/– p/p* will produce a rose comb *r/r P/–* will yield a pea comb.

In part a, a pea comb (*r/r p/p*) offspring was produced. Therefore, each parent had to possess a recessive allele at each of the two loci. Thus, the walnut parent had to be *R/r P/p*. The genotype of the single parent is given as *r/r p/p*.

	r p
R P	*R/r P/p*
R p	*R/r p/p*
r P	*r/r P/p*
r p	*r/r p/p*

In part b, a single comb offspring results, so we know that both parents had to each have the *p* allele. Similarly, in order to produce a pea offspring, each parent had to have the *r* allele. Hence the cross is *R/r p/p* (rose) x *R/r P/p* (walnut).

	R p	*r p*
R p	*R/R p/p* (rose)	*R/r p/p* (rose)
R P	*R/R R/p* (walnut)	*R/r P/p* (walnut)
r P	*R/r P/p* (walnut)	*r/r P/p* (pea)
r p	*R/r p/p* (rose)	*r/r p/p* (single)

In part c, the rose parent could be either *R/R p/p* or *R/r p/p*. The pea parent could be *r/r P/P* or *r/r P/p*. If both were homozygous for their dominant allele, all of the offspring would be walnut (*R/r P/p*). The ratio obtained is approximately 50 % walnut:50% rose. Hence, the

35

rose parent is the homozygote (*R/R p/p*) and the pea parent is heterozygous at the P locus (*r/r P/p*).

	r P	*r p*
R p	*R/r P/p*	*R/r p/p*
	(walnut)	(rose)

In part d, the fact that a single offspring (*r/r p/p*) was produced indicates that each walnut parent had to be a double heterozygote because each had to yield a recessive allele at both loci. The fact that no recognizable ratio results must be interpreted to mean that these parents had few offspring. (Perhaps they were someone's dinner at an early age!)

Answer: a. *R/r P/p* (walnut) x *r/r p/p* (single)
b. *R/r p/p* (rose) x *R/r P/p* (walnut)
c. *R/R p/p* (rose) x *r/r P/p* (pea)
d. *R/r P/p* (walnut) x *R/r P/p* (walnut)

4.17   In poultry feathered shanks (*F*) are dominant to clean (*f*), and white plumage of white leghorns (*I*) is dominant to black (*i*).
   a.   A feathered-shanked, white, rose-combed bird crossed with a clean-shanked, white, walnut-combed bird produces these offspring: two feathered, white, rose; four clean, white, walnut; three feathered, black, pea; one clean, black, single; one feathered, white, single; two clean, white, rose. What are the genotypes of the parents?
   b.   A feathered-shanked, white, walnut-combed bird crossed with a clean-shanked, white, pea-combed bird produces a single offspring, which is clean-shanked, black, and single-combed. In further offspring from this cross, what proportion may be expected to resemble each parent, respectively?

Answer: a. *F/f I/i R/r p/p* and *f/f I/i R/r P/p*
b. The parents are *F/f I/i R/r P/p* (A) and *f/f I/i r/r P/p* (B). The proportion of offspring phenotypically like parent A is (for the four gene pairs, respectively) 1/2 x 3/4 x 1/2 x 3/4 = 9/64. The proportion of offspring expected to be like parent B phenotypically is 1/2 x 3/4 x 1/2 x 3/4 = 9/64.

4.18 F_2 plants segregate 9/16 colored:7/16 colorless. If a colored plant from the F_2 is chosen at random and selfed, what is the probability that there will be no segregation of the two phenotypes among its progeny?

Answer: To show no segregation among the progeny the chosen plant must be homozygous. The genotypes comprising the 9/16 colored plants are: 1 *A/A B/B* :2 *A/a B/B*:2 *A/A B/b*:4 *A/a B/b*. Only one of these genotypes is homozygous; the answer is 1/9.

4.19   The gene *l* in *Drosophila* is a recessive, sex-linked gene, lethal when homozygous or hemizygous (the condition in the male). If a female of genotype *L/l* is crossed with a normal male, what is the probability that the first two surviving progeny to be observed will be males?

Answer: The cross is *L/l* x *L/l*. The progeny are 1/4 *L/L* ♀, 1/4 *L/l* ♀, 1/4 *L/Y* ♂ and 1/4 *l/Y* ♂. The *l/Y* die. Thus, the probability that a survivor will be male is 1/3. Therefore, the probability that the first two surviving progeny will be males is 1/3 x 1/3 = 1/9.

4.20 A locus in mice is involved with pigment production; when parents heterozygous at this locus are mated, 3/4 of the progeny are colored and 1/4 are albino. Another phenotype concerns the coat color produced in the mice; when two yellow mice are mated, 2/3 of the progeny are yellow and 1/3 are agouti. The albino mice cannot express whatever alleles they may have at the independently assorting agouti locus.

    a. When yellow mice are crossed with albino, they produce an F_1 consisting of 1/2 albino, 1/3 yellow, and 1/6 agouti. What are the probable genotypes of the parents?

    b. If yellow F_1 mice are crossed among themselves, what phenotypic ratio would you expect among the progeny? What proportion of the yellow progeny produced here would be expected to be true breeding?

Answer: a. If Y governs yellow and y governs agouti, then Y/y are lethal, Y/y are yellow, and y/y are agouti. Let C determine colored coat and c determine albino. The parental genotypes, then, are $Y/y\ C/c$ (yellow) and $Y/y\ c/c$ (white).

    b. The proportion is 2 yellow:1 agouti:1 albino. None of the yellows breed true since they are all heterozygous, with homozygous Y/Y individuals being lethal.

4.21 In *Drosophila melanogaster*, a recessive autosomal gene, ebony (e), produces black color when homozygous, and an independently assorting autosomal gene, black (b), also produces a black body color when homozygous. Flies with genotypes $e/e\ bl^+/-$, $e^+/-\ bl/bl$ and $e/e\ bl/bl$ are phenotypically identical with respect to body color. If true-breeding $e/e\ bl^+/bl^+$ ebony flies are crossed with true-breeding $e^+/e^+\ bl/bl$ black flies,

    a. what will be the phenotype of the F_1's?

    b. what phenotypes and what proportions would occur in the F_2 generation?

    c. what phenotypic ratios would you expect to find in the progeny of these backcrosses: (1) F_1 x true-breeding ebony and (2) F_1 x true-breeding black?

Answer: a. All are wild type: The flies are $+/e\ +/bl$ and the mutant alleles are recessive.

    b. The ratio is $9\ +\ +:3\ e\ +:3\ +\ bl:1\ e\ bl$, that is, 9 wild type:7 black body, since both mutants produce a black body color.

    c. (i) 1 wild type:1 ebony (ii) 1 wild type:1 black.

4.22 In four-o'clocks two genes, Y and R, affect flower color. Neither is completely dominant, and the two interact on each other to produce seven different flower colors:

        $Y/Y\ R/R$ = crimson $Y/y\ R/R$ = magenta
        $Y/Y\ R/r$ = orange-red $Y/y\ R/r$ = magenta-rose
        $Y/Y\ r/r$ = yellow $Y/y\ r/r$ = pale yellow
        $y/y\ R/R,\ y/y\ R/r,$ and $y/y\ r/r$ = white

    a. In a cross of a crimson-flowered plant with a white one ($y/y\ r/r$), what will be the appearances of the F_1 the F_2, and the offspring of the F_1 backcrossed to the crimson parent?

    b. What will be the flower colors in the offspring of a cross of orange-red x pale yellow?

    c. What will be the flower colors in the offspring of a cross of a yellow with a $y/y\ R/r$ white?

Answer: a. $Y/Y\ R/R$ (crimson) x $y/y\ r/r$ (white) gives $Y/y\ R/r$ F_1 plants, which have magenta-rose flowers. Selfing the F_1 gives an F_2 as follows: 1/16 crimson ($Y/Y\ R/R$), 2/16 orange-red ($Y/Y\ R/r$), 1/16 yellow ($Y/Y\ r/r$), 2/16 magenta ($Y/y\ R/R$), 4/16 magenta-rose ($Y/y\ R/r$), 2/16 pale yellow ($Y/y\ r/r$), and 4/16 white ($y/y\ R/R,\ y/y\ R/r,$ and $y/y\ rr$). Progeny of the F_1 backcrossed to the crimson parent are 1/4 crimson, 1/4 orange-red, 1/4 magenta, and 1/4 magenta-rose.

    b. $Y/Y\ R/r$ x $Y/y\ r/r$ gives 1/4 orange-red, 1/4 yellow, 1/4 magenta-rose, 1/4 pale yellow.

    c. $Y/Y\ r/r$ (yellow) x $y/y\ R/r$ (white) gives 1/2 magenta-rose ($Y/y\ R/r$) and 1/2 pale yellow ($Y/y\ r/r$).

4.23 Two four-o'clock plants were crossed and gave the following offspring: 1/8 crimson, 1/8 orange-red, 1/4 magenta, 1/4 magenta-rose, and 1/4 white. Unfortunately, the person who made the crosses was color-blind and could not record the flower colors of the parents. From the results of the cross, deduce the genotypes and flower colors of the two parents.

Solution: In the previous problem, we are told that the genes Y and R affect flower color in four-o'clocks. Neither is completely dominant and the interaction produces seven different colors as described in Question 4.22.

   To get any white offspring, each parent had to carry at least one y allele; to get a crimson flower, each parent had to carry at least one Y allele, hence each parent had to be heterozygous at the Y locus. Since there are no yellow or pale yellow offspring, it is unlikely that both parents were similarly heterozygous at the R locus. Indeed, the R locus assorts genotypically as R/R x R/r. Hence, the parents are magenta (Y/y R/R) x magenta-rose (Y/y R/r), and the cross is shown in the following Punnett square:

	$Y\,R$	$y\,R$
$Y\,R$	$Y/Y\ R/R$ (crimson)	$Y/y\ R/R$ (magenta)
$Y\,r$	$Y/Y\ R/r$ (orange-red)	$Y/y\ R/r$ (magenta-rose)
$y\,R$	$Y/y\ R/R$ (magenta)	$y/y\ R/R$ (white)
$y\,r$	$Y/y\ R/r$ (magenta-rose)	$y/y\ R/r$ (white)

Answer: Since individuals homozygous for y are found, both parents must have been heterozygous Yy. From the ratios of the phenotypes one parent must have been R/R and the other R/r. The parents therefore were Y/y R/R (magenta) and Y/y R/r (magenta-rose).

4.24 Genes A, B, and C are independently assorting and control production of a black pigment.
   a.   Assume that A, B, and C act in a pathway as follows:

$$\text{colorless} \quad \xrightarrow{A} \quad \xrightarrow{B} \quad \xrightarrow{C} \quad \text{black}$$

   The alternative alleles that give abnormal functioning of these genes are designated a, b, and c, respectively. A black $A/A\ B/B\ C/C$ is crossed by a colorless $a/a\ b/b\ c/c$ to give a black F_1. The F_1 is selfed. What proportion of the F_2 is colorless? (Assume that the products of each step except the last are colorless, so only colorless and black phenotypes are observed.)
   b.   Assume that C produces an inhibitor that prevents the formation of black by destroying the ability of B to carry out its function, as follows:

$$\text{colorless} \quad \xrightarrow{A} \quad \xrightarrow{B} \quad \text{black} \quad \rightarrow$$
$$\uparrow$$
$$C\ (\text{inhibitor})$$

A colorless A/A B/B C/C is crossed with a colorless a/a b/b c/c, giving a colorless F₁. The F₁ is selfed to give an F₂. What is the ratio of colorless to black in the F₂? (Only colorless and black phenotypes are observed, as in part a).

Answer: a. The simplest approach is to calculate the proportion of progeny that will be black and then subtract that answer from 1. The black progeny have the genotype $A/-B/-C/-$, and the proportion of these progeny is $(3/4)^3$. Therefore the proportion of colorless progeny is $1 - (3/4)^3 = 1 - 27/64 = 37/64$.

    b. Black is only produced when c/c $A/-B/-$ results, and this offspring occurs with the frequency $(1/4)(3/4)(3/4) = 9/64$ black, which gives $55/64$ colorless.

4.25 In *Drosophila* a mutant strain has plum-colored eyes. A cross between a plum-eyed male and a plum-eyed female gives 2/3 plum-eyed and 1/3 red-eyed (wild-type) progeny flies. A second mutant strain of *Drosophila*, called stubble, has short bristles instead of the normal long bristles. A cross between a stubble female and a stubble male gives 2/3 stubble and 1/3 normal-bristled flies in the offspring. Assuming that the plum gene assorts independently from the stubble gene, what will be the phenotypes and their relative proportions in the progeny of a cross between two plum-eyed, stubble-bristled flies? (Both genes are autosomal.)

Answer: The plum-eye allele is a dominant allele that gives plum eyes when heterozygous with the wild-type allele; it is lethal when homozygous. That is, Pm/Pm is lethal, Pm/Pm^+ gives plum eyes, and Pm^+/Pm^+ gives red eyes. A similar situation prevails for the stubble phenotype: Sb/Sb is lethal, Sb/Sb^+ gives stubble bristles, and Sb^+/Sb^+ gives normal bristles. The cross is Pm/Pm^+ Sb/Sb^+ x Pm/Pm^+ Sb/Sb^+, and the offspring are as follows:

1/16 Pm/Pm Sb/Sb	lethal
2/16 Pm/Pm Sb/Sb^+	lethal
1/16 Pm/Pm Sb^+/Sb^+	lethal
2/16 Pm/Pm^+ Sb/Sb	lethal
4/16 Pm/Pm^+ Sb/Sb^+	plum, stubble bristles
2/16 Pm/Pm^+ Sb^+/Sb^+	plum, normal bristles
1/16 Pm^+/Pm^+ Sb/Sb	lethal
2/16 Pm^+/Pm^+ Sb/Sb^+	red, stubble bristles
1/16 Pm^+/Pm^+ Sb^+/Sb^+	red, normal bristles

4.26 In sheep, white fleece (W) is dominant over black (w), and horned (H) is dominant over (h) in males but recessive in females. If a homozygous horned white ram is bred to a homozygous hornless black ewe, what will be the appearance of the F₁ and the F₂?

Answer: In males, H/H and H/h are horned, and h/h is hornless; in females H/H is horned, and H/h and h/h are hornless. The cross is a H/H W/W male x h/h w/w female. The F₁ is H/h W/w, which gives horned white males and hornless white females. Interbreeding the F₁ gives the following F₂:

		male	female
3/16	H/H $W/-$	horned, white	horned, white
6/16	H/h $W/-$	horned, white	hornless, white
3/16	h/h $W/-$	hornless, white	hornless, white
1/16	H/H w/w	horned, black	horned, black
2/16	H/h w/w	horned, black	hornless, black
1/16	h/h w/w	hornless, black	hornless, black

In sum, the ratios are 9/16 horned white:3/16 hornless white :3/16 horned black :1/16 hornless black males and 3/16 horned white:9/16 hornless white:1/16 horned black:3/16 hornless black females.

4.27   A horned black ram bred to a hornless white ewe has the following offspring: Of the males 1/4 are horned, white; 1/4 are horned, black; 1/4 are hornless, white; and 1/4 are hornless, black. Of the females 1/2 are hornless and black, and 1/2 are hornless and white. What are the genotypes of the parents?

Solution: The preceding question tells us that white fleece (*W*) is dominant to black (*w*), and horned (*H*) is dominant to hornless (*h*) in males but recessive in females.

Considering only the color of the fleece, the offspring of both sexes assort in a 50:50 ratio. Hence one of the parents had to be a heterozygote and the other hornozygous for the black color. Similarly, the gene for horns assorts in a 50:50 ratio, again suggesting that one parent is heterozygous and the other homozygous for this trait. The distinction must be then made between these crosses:

1) *W/w H/h* x *w/w h/h*   or   2) *W/w h/h* x *w/w H/h*

Since the problem specifies a black, horned ram., cross #1 can not be correct since neither of those genotypes would translate to a black, hornless ram. Cross #2 could either be a white, hornless ewe x a black, horned ram, or a white, hornless ram x a black, hornless ewe.

Let us try Cross #2, keeping in mind that the genotypes *H/h* and *H/H* produce horns in males, but to get horns in females, one would require only *H/H*. (With sex-influencee, traits such as this one, it may be less confusing to use symbols other than upper and lower case letters. For example, H_1 and H_2 where H_1 is dominant in males and H_2 is dominant in females.)

W/w h/h female × w/w H/h male

	W h	w h
w H	W/w H/h white, horned male white, hornless female	w/w H/h black, horned male black, hornless female
w h	W/w h/h white, hornless male white, hornless female	w/w h/h black, hornless male black, hornless female

Answer: The *H/h w/w* male x *h/h W/w* female gives the following results:

		male	female
1/4	H/h W/w	horned, white	hornless, white
1/4	h/h W/w	hornless, white	hornless, white
1/4	H/h w/w	horned, black	hornless, black
1/4	h/h w/w	hornless, black	hornless, black

In sum, the ratios are 1/4 horned, white:1/4 hornless, white:1/4 horned, black:1/4 hornless, black males and 1/2 hornless, white:1/2 hornless, black females.

4.28 A horned white ram is bred to the following four ewes and has one offspring by the first three and two by the fourth: Ewe A is hornless and black; the offspring is a horned white female. Ewe B is hornless and white; the offspring is a hornless black female. Ewe C is horned and black; the offspring is a horned white female. Ewe D is hornless and white; the offspring are one hornless black male and one horned white female. What are the genotypes of the five parents?

Answer: Ewe A is *H/h w/w*; B is *H/h W/w* or *h/h W/w*; C is *H/H w/w*; D is *H/h W/w*; the ram is *H/h W/w*.

4.29 Common pattern baldness is more frequent in males than in females. This appreciable difference in frequency is assumed to be due to
   a. Y-linkage of this trait.
   b. X-linked recessive mode of inheritance for the trait in question.
   c. sex-influenced autosomal inheritance.
   d. excessive beer-drinking in males, consumption of gin being approximately equal between the sexes.

Answer: c

CHAPTER 5

LINKAGE, CROSSING-OVER, AND GENE MAPPING IN EUKARYOTES

I. CHAPTER OUTLINE

Discovery of Genetic Linkage
    Morgan's Linkage Experiments with *Drosophila*
Gene Recombination and the Role of Chromosomal Exchange
    Corn Experiments
    Crossing-Over at the Tetrad (Four-Chromatid) Stage of Meiosis
Locating Genes on Chromosomes: Mapping Techniques
    Detecting Linkage through Testcrosses
    Gene Mapping Using Two-Point Testcrosses
    Generating a Genetic Map
    Double Crossovers
    Three-Point Cross
    Mapping Chromosomes by Using Three-Point Testcrosses

II. IMPORTANT TERMS AND CONCEPTS

linked genes	coupling *vs.* repulsion
linkage groups	map unit
linkage (genetic) map	centi-Morgan
partial linkage	multiple crossovers
parentals *vs.* recombinants	double crossover
cytological *vs.* genetic marker	mapping functions
ordered tetrads	chiasma (chromosomal) interference
chi-square test	coefficient of coincidence

III. THINKING ANALYTICALLY

When you first begin to work mapping problems, remember that <u>organization</u> counts. Do not try to solve problems without reorganizing the data so that you can extract several important pieces of information by inspection. Thus, reorder the progeny of a three point cross into pairs of reciprocals. List the most numerous pair first - these will be the parentals or nonrecombinants. List the least numerous last - these will be the double crossovers. From a comparison of these two groups, you can deduce which locus is in the middle.

Work consistently with percentages or decimals, but do not mix the two. Remember that when calculating coefficient of coincidence (or interference) the ratio should be structured with decimals in order to get a correct answer in percent. Do not attempt the reverse.

Before progressing in mapping problems, first be sure that the genes are linked. Remember that the rule for linked genes (less than 50% recombination) applies to a test-cross (hybrid crossed with homozygous recessive) and not to a typical cross between F_1 individuals.

The same advice regarding organization and consistency applies to chi-square tests. When interpreting chi-square datal remember that, like most statistical tests, this one serves as a guide for invalidation, not validation. Review the text until you fully understand what is being summed to obtain the chi-square valuer and what the P, or probability, value is really measuring.

IV. QUESTIONS FOR PRACTICE

A. Multiple Choice Questions

1.   The significance of ordered tetrads in *Neurospora* was to show that crossing-over
        a.   occurs at the 2 chromatid stage.
        b.   occurs at the 4 chromatid stage.
        c.   can occur when meiosis does not occur.
        d.   only occurs when meiosis occurs.
        e.   occurs after meiosis has been completed.

2.   Alleles are considered to be linked if recombination is less than 50% in a (an)
        a. $P \times P \rightarrow F_1$ cross.
        b. $F_1 \times F_1 \rightarrow F_2$ cross.
        c. $F_2 \times F_2 \rightarrow F$ cross.
        d. $F_1 \times$ homozygous recessive cross.
        e.   any F_2 progeny x homozygous recessive cross.

3.   In the cross of a blood-type A individual with a blood-type B individual, progeny of all four blood types were obtained. If you were doing a chi-square test on such a cross how many degrees of freedom would you use?
        a.   1
        b.   2
        c.   3
        d.   4
        e.   0

4.   The chi-square test is best used to decide whether an hypothesis is
        a.   logical.
        b.   unacceptable.
        c.   temporary.
        d.   permanent.
        e.   correct.

5.   A chi-square value of 0.95 at 1 degree of freedom means that
        a.   95% of the data are correct.
        b.   the hypothesis is probably incorrect.
        c.   your hypothesis has a 30-50% chance of being correct.
        d.   your hypothesis has a 95% chance of being correct.
        e.   between 30 to 50% of the time, chance deviations as great or greater than those obtained in this experiment would be obtained.

Answers: 1b, 2d, 3a, 4b, 5e

B. Thought Questions

1.   Do you think that mapping according to the methods described in this chapter will survive in the future when DNA sequencing will be commonplace? (Hint: Does genetic and/or cytological mapping narrow the field of possibilities?)

2. Why do you think that sex linkage was described before other examples of linkage in animals? (Hint: Is this an example of a correlation of genetic and cytologic data?)

3. Why were the correlations made by Creighton and McClintock in their study of the waxy and colorless loci in corn important? (Hint: What was the value of cytological data to genetic experiments in the early days of this century?)

4. Can one make genetic maps of haploid organisms? (Hint: review *Neurospora* material.)

5. Does the decision to accept or reject an hypothesis at the 0.05 level mean that one is more likely to discard a correct hypothesis or retain an incorrect one? (Hint: Review the chi-square material and think about the long-term consequences of these alternatives.)

V. ANSWERS AND SOLUTIONS TO TEXT QUESTIONS

5.1 A cross $a^+a^+ b^+b^+$ x aa bb results in an F_1 of phenotype a^+b^+; the following numbers are obtained in the F_2 (phenotypes):

a^+	b^+	110
a^+	b	16
a	b^+	19
a	b	15
Total		160

Are genes at the a and b loci linked or independent? What F_2 numbers would otherwise be expected?

Answer: They are linked; if they were independent, the expected numbers would be 90:30:30:10.

5.2 In the F_2 of his cross of red-flowered x white-flowered *Pisum*, Mendel obtained 705 plants with red flowers and 224 with white.
   a. Is this result consistent with his hypothesis of factor segregation, from which a 3:1 ratio would be predicted?
   b. In how many similar experiments would a deviation as great as or greater than this one be expected? (Calculate χ^2 and obtain the approximate value of P from the table.)

Answer:

Class	observed	expected	d	d^2	d^2/e
red	705	697	8	64	0.09
white	224	232	- 8	64	0.27
Total	929	929	0	–	$0.36 = \chi^2$

$\chi^2 = 0.36$; df =1; $P \approx 0.60$.

   a. The result is consistent with the hypothesis.
   b. In approximately 60% of similar experiments we could expect a deviation as great as or greater than this one.

5.3 In corn a dihybrid for the recessives a and b is testcrossed. The distribution of the phenotypes was as follows:

A	B	122
A	b	118
a	B	81
a	b	79

Are the genes assorting independently? Test the hypothesis with a χ^2 test. Explain tentatively any deviation from expectation, and tell how you would test you explanation.

Answer: From the χ^2 test, $\chi^2 = 16.10$; P is less than 0.01 at three degrees of freedom. This test reveals that the two genes do not fit a 1:1:1:1 ratio. It does not say why. Linkage might seem reasonable until it is realized that the minority classes are not reciprocal classes (both carry the *aa* phenotype). If the segregation at each locus is considered, however, the *B/–:b/b* ratio is about 1:1 (203:197), while the *A/–:a/a* ratio is not (240:160). The departure, then, is specifically the result of a deficiency of *a/a* individuals. This departure should be confirmed in other crosses that test the segregation at locus *A*. In corn further evidence would be that the *a/a* deficiency might show up as a class of ungermininated seeds or of seedlings that die early.

5.4 Why are crosses used for chromosome mapping *Drosophila* set up with females heterozygous and males homozygous recessive?

Answer: Because there is no crossing over in male *Drosophila*.

5.5 The F_1 from a cross of *A B/A B* x *a b/a b* is testcrossed, resulting in the following phenotypic ratios:

A	B	308
A	b	190
a	b	292
a	B	210

What is the frequency of recombination between genes *a* and *b*?

Answer: 400/1000 x 100 = 40%

5.6 In rabbits the English type of coat (white-spotted) is dominant over non-English (unspotted), and short hair is dominant over long hair (Angora). When homozygous English, short-haired rabbits were crossed with non-English Angoras and the F_1 crossed back to non-English Angoras, the following offspring were obtained: 72 English and short-haired, 69 non-English and Angora; 11 English and Angora; and 6 non-English and short-haired. What is the map distance between the genes for coat color and hair length?

Answer: There are 158 progeny rabbits. The recombinant classes are the English plus Angora and the non-English plus short-haired, so the map distance between the genes is [(11 + 6)/158] x 100% = 10.8% = 10.8 mu.

5.7 In *Drosophila* the mutant black (*b*) has a black body, and the wild type has a grey body; the mutant vestigial (*vg*) has wings that are much shortened and crumpled when compared with the long wings of the wild type. In the following cross, the true-breeding parents are given together with the counts of offspring of F_1 females x black and vestigial males:

P black and normal x grey and vestigial
F_1 females x black and vestigial males

grey, normal	283
grey, vestigial	1294
black, normal	1418
black, vestigial	241

From these data, calculate the map distance between the black and vestigal genes.

Answer: The cross is *b vg$^+$/b vg$^+$* x *b$^+$ vg/b$^+$ vg*, which gives a *b$^+$ vg/b vg$^+$* F_1. An F_1 female is crossed with a homozygous recessive *b vg/b vg*. The female is used as the doubly heterozygous parent since no crossing-over occurs in male *Drosophila*. The classes can be grouped in reciprocal pairs as shown on the following page:

Non-recombinants/parentals:	grey, vestigial	1294
	black, normal	1418
Recombinants/non-parentals:	grey, normal	283
	black, vestigial	241
	Total progeny	3236

Total non-recombinants: 1294 + 1418 = 2712
Total recombinants: 283 + 241 = 524
% recombinants = (524/3236) x 100% = 16.2.
Therefore the map distance is 16.2 map units.

5.8   A gene controlling wing size is located on chromosome 2 in *Drosophila*. The recessive allele *vg* results in vestigial wings when homozygous; the vg^+ allele determines long wings. A new eye mutation, which we will call "maroon-like," is isolated. Homozygous maroon-like (*m/m*) results in maroon colored eyes; the m^+ allele is bright red. The location of the *m* gene is unknown, and you are asked to design an experiment to determine whether *m* is located on chromosome 2.
    You cross true-breeding virgin maroon females to true-breeding *vg/vg* males and obtain all wild-type F_1 progeny. Then you allow the F_1 to interbreed. As soon as the F_2 start to hatch, you begin to classify the flies, and among the first six newly-hatched flies, you find four wild type; one vestigial-winged and red-eyed fly; and one vestigial-winged, maroon-eyed fly. You immediately draw the conclusions that (1) maroon-like is not X-linked and (2) maroon-like is not linked to vestigial. Based on this small sample, how could you tell? On what chromosome is *m* located? (Hint: There is no crossing-over in *Drosophila* males.)

Answer: You were lucky in that a double mutant hatched. Based on the presence of one double mutant, you can eliminate linkage between *vg* and *m*. This is because there is no crossing-over in the *Drosophila* male, and there is no way to produce a recombinant *vg m* gamete in the male F_1. The crosses performed here were

$$P \quad \frac{vg^+}{vg^+} \quad \frac{m}{m} \; \female \quad \times \quad \frac{vg}{vg} \quad \frac{m^+}{m^+} \; \male$$

$$\downarrow$$

All F_1 are $\quad \frac{vg^+}{vg} ; \frac{m^+}{m} \quad$ (wild type)

The results of this cross told you that *m* is not x-linked.
    Therefore the F_1 cross was a dihybrid cross and we expect a 9:3:3:1 F_2 (wild type:*vg* m^+:vg^+ *m*:*vg m*) ratio. In the small sample, by chance, a double mutant appeared. This tells us tht *vg* and *m* are segregating independently. If *vg* and *m* had been on the same chromosome, the crosses performed would have been

$$P \quad \frac{vg^+ \quad m}{vg^+ \quad m} \quad \times \quad \frac{vg \quad m^+}{vg \quad m^+}$$

$$\downarrow$$

$$F_1 \quad \frac{vg^+ \quad m}{vg \quad m^+} \quad \text{(All wild type)}$$

♂ gametes ♀ gametes	vg^+ m	vg m^+
vg^+ m	maroon	wild type
vg m^+	wild type	vestigial
vg^+ m^+	wild type	wild type
vg m	maroon	vestigial

The phenotypic ratio would be 4:1:1:0 (wild:vestigial:maroon:vestigial maroon).

There is no way to get a double mutant from this cross <u>regardless</u> of the distance between the two linked genes (0-100 percent recombination). So if the two genes were linked one would never find a double mutant progeny in the F_2. Since a double mutant did hatch, this was enough to allow the preliminary (but correct) conclusion that m was not on chromosome 2. While m could be on chromosome 3 or 4, this experiment does not distinguish between these two possiblities.

5.9   Use the following two-point data to map the genes concerned. Show the order and the length of the shortest intervals.

Gene Loci	% Recom- bination	Gene Loci	% Recombination
a,b	50	b,d	13
a,c	15	b,e	50
a,d	38	c,d	50
a,e	8	c,e	7
b,c	50	d,c	45

Answer:

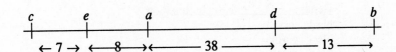

Note that the map distances are not strictly additive - e.g. *a - d* = 38 and *a - e* = 8 but *d - e* = 45 - because of the effects of multiple crossovers, as described in the chapter.

5.10 Use the following two-point recombination data to map the genes concerned. Show the order and the length of the shortest intervals.

Loci	% Recombination	Loci	% Recombination
a,b	50	c,d	50
a,c	17	c,e	50
a,d	50	c,f	7
a,e	50	c,g	19
a,f	12	d,e	7
a,g	3	d,f	50
b,c	50	d,g	50
b,d	2	e,f	50
b,e	5	e,g	50
b,f	50	f,g	15
b,g	50		

Answer:

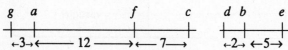

5.11 The following data are from Bridges and Morgan's work on the recombination betweeen the genes black, curved, purple, speck, and vestigial in chromosome 2 of *Drosophila*. On the basis of the data map the chromosome for these five genes as accurately as possible. Remember that determinations for short distances are more accurate than those for long ones.

Genes in cross	Total progeny	Number of recombinants
black, curved	62,679	14,237
black, purple	48,931	3,026
black, speck	685	326
black, vestigial	20,153	3,578
curved, purple	51,136	10,205
curved, speck	10,042	3,037
curved, vestigial	1,720	141
purple, speck	11,985	5,474
purple, vestigial	13,601	1,609
speck, vestigial	2,054	738

Solution: First determine that the total number of progeny is 62,679. Divide the number of recombinants by that total to determine the % recombination between each pair of genes.

Genes	% Recombination
black, curved	22.7
black, purple	6.18
black, speck	47.6
black, vestigial	17.8
curved, purple	20.0
curved, speck	30.2
curved, vestigial	8.2
purple, speck	45.7
purple, vestigial	11.8
speck, vestigial	35.9

The pair with the greatest frequency of recombination (black, speck - 47.6%) is probably the farthest apart. Draw the map with these genes marking the ends:

Next fill in the order by finding the sequence of genes that has an increasing order of recombination with black:

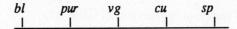

Next assign distances on your map using the recombination frquencies between neighboring genes:

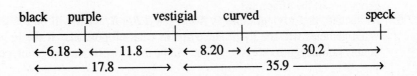

Check your results for gene <u>order</u> by using <u>curved</u> as the point of reference; if you wish you could pick each of the genes in turn as a reference point. Each time you check, look for increasing frequencies of recombination to determine linear order. Finally, add the intermediate distances and compare them to the more inclusive distances; the sum of two smaller distances should always be the larger. The larger distances are usually too low because multiple crossovers have hidden some percentage of the recombinant events. Thus, the distance given for the black-speck recombinants is 47.6 map units, but the total of all of the smaller distances is 56.4. In summary, the map (not to scale) will look like:

Answer:

```
   black   purple        vestigial   curved                    speck
   +-------+-------------+-----------+------------------------+
    ←6.18→ ←—— 11.8 ——→ ← 8.20 → ←———— 30.2 ————→
           ←———— 17.8 ———→ ←————— 35.9 —————→
```

5.12   Genes *a* and *b* are linked, with 10 percent recombination. What would be the phenotypes, and the probability of each, among progeny of the following cross?

$$\frac{a \quad b^+}{a^+ \quad b} \quad \text{x} \quad \frac{a \quad b}{a \quad b}$$

Answer: 45% *a b*$^+$, 45% *a*$^+$ *b*, 5% *a*$^+$ *b*$^+$, 5% *a b*

5.13   Genes *a* and *b* are sex-linked and are located 7 mu apart in the X chromosome of *Drosophila*. A female of genotype *a*$^+$ *b*/*a b*$^+$ is mated with a wild-type (*a*$^+$ *b*$^+$) male.
   a.   What is the probability that one of her sons will be either *a*$^+$ *b*$^+$ or *a b*$^+$ in phenotype?
   b.   What is the probability that one of her daughters will be *a*$^+$ *b*$^+$ in phenotype?

Answer: a.   0.035 + 0.465 = 0.50
           b.   All of the daughters have *a*$^+$ *b*$^+$ phenotype.

5.14   *a* and *b* are linked autosomal genes in *Drosophila* whose recombination frequency in females is 5 percent; *c* and *d* are X-linked genes, located 10 map units apart. A homozygous dominant female is mated to a recessive male and the daughters are testcrossed. Which of the following would you expect to observe in the testcross progeny?
   a.   Different ratios in males and females.
   b.   Nearly equal frequency of *a*$^+$ *b c*$^+$ *d*$^+$, *a*$^+$ *b c d*, *a b*$^+$ *c*$^+$ *d*$^+$, and *a b*$^+$ *c d* classes.
   c.   Independent segregation of some genes with respect to others involved in the cross.
   d.   Double-crossover classes less frequent than expected because of in the two marked regions.

Answer: F₁ daughters are

$$\frac{a^+ \ b}{a \ \ b} \quad \frac{c^+ \ d^+}{c \ \ d}$$

Would expect (b) and (c) to be true, but not (a) or (d).

5.15 In maize the dominant genes *A* and *C* are both necessary for colored seeds. Homozygous recessive plants give colorless seed, regardless of the genes at the second locus. Genes *A* and *C* show independent segregation, while the recessive mutant gene waxy endosperm (*wx*) is linked with *C* (20 percent recombination). The dominant *Wx* allele results in starchy endosperm.
   a. What phenotypic ratios would be expected when a plant of constitution *c Wx/C wx A/A* is testcrossed?
   b. What phenotypic ratios would be expected when a plant of constitution *c Wx/C wx A/a* is testcrossed?

Answer: a. 40% colorless, starchy; 40% colored, waxy; 10% colorless, waxy; 10% colored, starchy
   b. 45% colorless, starchy; 20% colored, waxy; 30% colorless, waxy; 5% colored, starchy

5.16 Assume that genes *a* and *b* are linked and show 20 percent crossing-over.
   a. If a homozygous *A B/A B* individual is crossed with a *a b/a b* individual, what will be the genotype of the F₁? What gametes will the F₁ produce and in what proportions? If the F₁ is testcrossed with a doubly homozygous recessive individual, what will be the proportions and genotypes of the offspring?
   b. If, instead, the original cross is *A b/A b* x *a B/a B*, what will be the genotype of the F₁? What gametes will the F₁ produce and in what proportions? If the F₁ is testcrossed with a doubly homozygous recessive, what will be the proportions and genotypes of the offspring?

Answer: a. The genotype of the F₁ is *A B/a b*. The gametes produced by the F₁ are 40% *A B*; 40% *a b*; 10% *A b*; 10% *a B*. From a testcross of the F₁ with *a b/a b* we get 40% *A B/a b*; 40% *a b/a b*; 10% *A b/a b*; 10% *a b/a b*.
   b. The F₁ genotype is *A b/a B*. The gametes produced by the F₁ are 40% *A b*; 40% *a B*; 10% *A B*; 10% *a b*. From a testcross of the F₁ with *a b/a b* we get 40% *A b/a b*; 40% *a b/a b*; 10% *A B/a b*; 10% *a b/a b*. This question illustrates that map distance is computed between sites in the chromosome, and it does not matter whether the genes are in coupling or in repulsion, the percentage of recombinants will be the same, even though the particular phenotypic classes constituting the recombinants differ.

5.17 In tomatoes, tall vine is dominant over dwarf, and spherical fruit shape is dominant over pear shape. Vine height and fruit shape are linked, with a recombinant percentage of 20. A certain tall, spherical-fruited tomato plant is crossed with a dwarf, pear-fruited plant. The progeny are 81 tall, spherical; 79 dwarf, pear; 22 tall, pear; and 17 dwarf, spherical. Another tall and spherical plant crossed with a dwarf and pear plant produces 21 tall, pear, 18 dwarf, spherical; 5 tall, spherical; and 4 dwarf, pear. What are the genotypes of the two tall and spherical plants? If they were crossed, what would their offspring be?

Answer: In the first cross four classes of progeny were produced with about 20% tall plus pear and dwarf plus spherical progeny. Thus the tall, spherical parent was doubly heterozygous, and the genes were in coupling, that is, *D P/d p*, for *D* tall, *d* dwarf, *P* spherical and *p* pear. In the other cross there are also four classes of progeny, with about 20% tall plus spherical and dwarf plus pear progeny. Thus the tall spherical plant was also doubly heterosygous with the genes this time in repulsion, for example, *D p/d P*. The cross between the two tall, is *D P/d p* x *D p/d P*. The offspring produced and their phenotypes are in the following figure:

$$D\,P/d\,p \quad \times \quad D\,p/d\,P$$

tall, spherical ↓ tall, spherical

	$D\,p$ 0.4	$d\,P$ 0.4	$D\,P$ 0.1	$d\,p$ 0.1
$D\,P$ 0.4	$D\,P/D\,p$ tall, spherical 0.16	$D\,P/d\,P$ tall, spherical 0.16	$D\,P/D\,P$ tall, spherical 0.04	$D\,P/d\,p$ tall, spherical 0.04
$d\,p$ 0.4	$D\,p/d\,p$ tall, pear 0.16	$d\,p/d\,P$ dwarf, spherical 0.16	$d\,p/D\,P$ tall, spherical 0.04	$d\,p/d\,p$ dwarf, pear 0.04
$D\,p$ 0.1	$D\,p/D\,p$ tall, pear 0.04	$D\,p/d\,P$ tall, spherical 0.04	$D\,p/D\,P$ tall, spherical 0.01	$D\,p/d\,p$ tall, pear 0.01
$d\,P$ 0.1	$d\,P/D\,p$ tall, spherical 0.04	$d\,P/d\,P$ dwarf, spherical 0.04	$d\,P/D\,P$ tall, spherical 0.01	$d\,P/d\,p$ dwarf, spherical 0.01

Phenotypes:

tall, spherical → 0.16 + 0.16 + 0.04 + 0.04 + 0.04 + 0.04 + 0.01 + 0.04 + 0.01 = *0.54*

tall, pear → 0.16 + 0.04 + 0.01 = *0.21*

dwarf, spherical → 0.16 + 0.04 + 0.01 = *0.21*

dwarf, pear → *0.04*

5.18 Genes *a* and *b* are in one chromosome, 20 mu apart; *c* and *d* are in another chromosome, 10 mu apart. Genes *e* and *f* are in yet another chromosome and are 30 mu apart. Cross a homozygous *A B C D E F* individual with an *a b c d e f* one, and cross the F₁ back to an *a b c d e f* individual. What are the chances of getting individuals of the following phenotypes in the progeny?
   a. *A B C D E F*
   b. *A B C d e f*
   c. *A b c D E*
   d. *a B C d e f*
   e. *a b c D e F*

Answer: Each chromosome pair segregates independently. We can compute the relative proportions of gametes produced for each homologous pair of chromosomes separately from the known map distances (P = parental; R = recombinant):

P		R		P		R		P		R	
Ab	0.4	Ab	0.1	CD	0.45	Cd	0.05	EF	0.35	Ef	0.15
ab	0.4	aB	0.1	cd	0.45	cD	0.05	ef	0.35	eF	0.15

To answer the questions, simply multiply the probabilities of getting the particular gamete from the F_1 multiple heterozygote.

    a. *A B C D E F* = 0.4 x 0.45 x 0.35 = 0.063 (6.3%)
    b. *A B C d e f* = 0.4 x 0.05 x 0.35 = 0.007 (0.7%)
    c. *A b c D Ef* = 0.1 x 0.05 x 0.15 = 0.00075 (0.075%)
    d. *a B C d e f* = 0.1 x 0.05 x 0.35 = 0.00175 (0.175%)
    e. *a b c D e F* = 0.4 x 0.05 x 0.15 = 0.003 (0.3%)

5.19   Genes *d* and *p* occupy loci 5 map units apart in the same autosomal linkage group. Gene *h* is a separate autosomal linkage group and therefore segregates independently of the other two. What types of offspring are expected, and what is the probability of each, when individuals of the following genotypes are testcrossed:

    a. $\dfrac{D\ P}{d\ p}$ $\dfrac{h}{h}$ b. $\dfrac{D\ P}{D\ p}$ $\dfrac{h}{h}$

Answer: a. 47.5% each of *D P h* and *d p h*;;2.5% each of *D p h* and *d P h*.
    b. 23.75% each of *D p H, D p h, d P H,* and *d P h*; 1.25% each of *D P H, D P h, d p H,* and *d p h*.

5.20   A hairy-winged (*h*) *Drosophila* female is mated with a yellow-bodied (*y*), white-eyed (*w*) male. The F_1 are all normal. The F_1 progeny are then crossed, and the F_2 that emerge are as follows:

Females:	wild type	757
	hairy	243
Males:	wild type	390
	hairy	130
	yellow	4
	white	3
	hairy, yellow	1
	hairy, white	2
	yellow, white	60
	hairy, yellow, white	110

Give genotypes of the parents and the F_1, and note the linkage relations and distances, where appropriate.

Answer: Parents: The male is h^+/h^+; *y w*/Y (autosomal and XY pair, respectively); the female is *h/h*; y^+w/y^+ *w*. The F_1 male is h^+/h; y^+ w^+/Y; the female is h^+/h; y^+ w^+/y *w*. Linkage relation: *h* is on an autosome, unlinked to *y* and *w*; *y* and *w* are on the X chromosome, 10/1000 = 1 mu apart (Note that the ratio of hairy:wild is 1:3 in males also.)

5.21   In the Maltese bippy amiable (*A*) is dominant to nasty (*a*), benign (*B*) is dominant to active (*b*), and crazy (*C*) is dominant to sane (*c*). A true-breeding amiable, active, crazy bippy was mated, with some difficulty, to a true-breeding nasty, benign, sane bippy. An F_1 individual from this cross was then used in a testcross (to a nasty, active, sane bippy) and produced, in typical prolific bippy fashion, 4000 offspring. From an ancient manuscript entitled *The Genetics of the Bippy,*

Maltese and Other, you discover that all three genes are autosomal, *a* is linked to *b* but not to *c* and the map distance between *a* and *b* is 20 mu.

  a. Predict all the expected phenotypes and the numbers of each type from this cross.
  b. Which phenotypic classes would be missing had *a* and *b* shown complete linkage?
  c. Which phenotypic classes would be missing if *a* and *b* were unlinked?
  d. Again, assuming *a* and *b* to be unlinked, predict all the expected phenotypes of nasty bippies and the frequencies of each type resulting from a self-cross of the F_1.

Answer: a. *A* is linked to *B*, but neither is linked to *C*. The cross involves *A* and *B* in repulsion:

$$\frac{A\ b}{A\ b}\quad \frac{C}{C}\quad \times \quad \frac{a\ B}{a\ B}\quad \frac{c}{c}$$

This cross gives the following F_1:

$$\frac{A\ b}{a\ B}\quad \frac{C}{c}$$

The F_1 is crossed with a triply homozygous recessive. There are 20 mu between *A* and *B*, so the total of all recombinants involving *A* and *B* add up to 20% of all progeny. The progeny, and their proportions, are as follows:

Parentals:				
	A b C	amiable, active, crazy	800	(20%)
	A b c	amiable, active, sane	800	(20%)
	a B C	nasty, benign, crazy	800	(20%)
	a B c	nasty, benign, sane	800	(20%)
Recombinants:				
	A B C	amiable, benign, crazy	200	(5%)
	A B c	amiable, benign, sane	200	(5%)
	a b C	nasty, active, crazy	200	(5%)
	a b c	nasty, active, sane	200	(5%)

  b. If *A* and *B* were completely linked, none of the recombinant classes would be produced.
  c. If *A* and *B* were unlinked, none of the classes would be missing, although there would be equal numbers then of each of the *B* phenotypic classes, that is, 500 each.
  d. If all three genes are unlinked, the results of selfing a triply heterozygote, the F_1 are as described for a trihybrid cross in Chapter 2. That is, in the F_2 there are B phenotypic classes in a ratio of 27:9:9:9:3:3:3:1. The nasty bippies must be an *aa* in genotype, so the distribution is essentially a subset of the previous classes, that is, 9 nasty, benign, crazy:3 nasty, benign, sane:3 nasty, active, crazy:1 nasty, active, sane.

5.22 How many possible linear orders are there for three linked genes? How many are there if the left versus the right end of the chromosome is taken into consideration?

Answer: 3; 6

53

5.23 Fill in the blanks. Continuous bars indicate linkage, and the order of linked genes is correct as shown. If all types of gametes are equally probable, write "none" in the right column headed "Least frequent classes." In the right column, show two gamete genotypes, unless all types are equally frequent, in which case write "none."

Parent genotypes	Number of different possible gamete	Least frequent classes
$\underline{A\ b\ C}$ $a\ B\ c$	_____	_____ _____
$\underline{A\ b\ C}$ $a\ B\ c$	_____	_____ _____
$\underline{A\ b\ C\ D}$ $a\ B\ c\ d$	_____	_____ _____
$\underline{A\ b\ C\ D\ E\ f}$ $a\ B\ C\ d\ e\ f$	_____	_____ _____
$\underline{b\ D}$ $B\ d$	_____	_____ _____

Answer: The following answers are the number of different possible gamete genotypes and the least frequent classes given respectively for the five parent genotypes given:

2^3 none

2^3 $A\ B\ C$ and $a\ b\ c$

2^4 $A\ B\ C\ d$ and $a\ b\ c\ D$

2^2 $A\ b\ C\ d\ e\ f$ and $A\ B\ C\ D\ e\ f$

2^2 $b\ d$ and $B\ D$

5.24 For each of the following tabulations of testcross progeny phenotypes and numbers, state which locus is in the middle, and reconstruct the genotype of the tested triple heterozygotes.

a.
A B C	191
a b c	180
A b c	5
a B C	5
A B c	21
a b C	31
A b C	104
a B c	109

b.
C D E	9
c d e	11
C d e	35
c D E	27
C D e	78
c d E	81
C d E	275
c D e	256

c.
F G H	110
f g h	114
F g h	37
f G H	33
F G h	202
f g H	185
F g H	4
f G h	0

Answer:
a. *a* is in the middle. The genotype of the triple heterzygote is $B\ A\ C/b\ a\ c$.
b. *d* is in the middle. The genotype is $C\ d\ E/c\ D\ e$.
c. *f* is in the middle. The genotype is $G\ F\ h/g\ f\ H$.

5.25   Genes at loci *f*, *m*, and *w* are linked, but their order is unknown. The F_1 heterozygotes from a cross of *FF MM WW* x *ff mm ww* are testcrossed. The most frequent phenotypes in testcross progeny will be *F M W* and *f m w* regardless of what the gene order turns out to be.

      a.   What classes of testcross progeny (phenotypes) would be least frequent if locus *m* is in the middle?

      b.   What classes would be least frequent if locus *f* is in the middle?

      c.   What classes would be least freqent if locus *w* is in the middle?

Answer: a. *F m W* and *f M w* will be the lease frequent (because of double crossovers).

              b. *M f W* and *m F w* will be the least frequent

              c. *M w F* and *m W f* will be the least frequent

5.26   The following numbers were obtained for testcross progeny in *Drosophila* (phenotypes):

+	*m*	+	218
w	+	*f*	236
+	+	*f*	168
w	*m*	+	178
+	*m*	*f*	95
w	+	+	101
+	+	+	3
w	*m*	*f*	1
		Total	1000

Construct a genetic map.

Solution: Notice that the data are already listed in reciprocal pairs. The most numerous pair, presumably the nonrecombinants, heads the list while the double crossovers form the tail of the list. The problem states that the cross was a testcross and that the progeny are listed by phenotype. From this information, the cross must have been between triple heterozygotes and homozygous recessives. The members of the P generation had to have been homozygous for all loci, and from inspection of the non-recombinant classes, they must have been:

    P   _+ m +/+ m +_   x   _w + f/w + f_

                                  ↓

    F_1 _+ m +/w + f_

When the F_1 is testcrossed, the phenotype of the progeny will reflect the genotype of the F_1 parent.

    Next, construct a table of recombination in which the genes are considered in pairs, and score each event in which the progeny show any recombination when compared to the order to alleles in the F_1 organism. If recombination is noted between two of the three pairs, then enter the number of progeny in each appropriate column, as follows:

				Recombination between			
				w and *m*	*m* and *f*	*w* and *f*	
+	*m*	+	218	0	0	0	parental
w	+	*f*	236	0	0	0	parental
+	+	+	168	168	0	168	
w	*m*	+	178	178	0	178	
+	*m*	*f*	95	0	95	95	
w	+	+	101	0	101	101	
+	+	+	3	3	3	0	
w	*m*	*f*	1	1	1	0	
Totals			1000	346	200	542	

Recombination between w and m is 346/1000 = 34.6%
Recombination between m and f is 200/1000 = 20%
Recombination between w and f is 542/1000 = 54.2%

The map must show w and f as the terminals, with m in the middle. Alternatively, one could deduce the order by comparing the smallest recombinant classes (+ + + and w m f) with that of the parentals, and figuring what order would be consistent with a double crossover event. In this problem, the genes are listed in their correct order, but don't expect that all of the time! Since smaller distances are more accurate than larger ones, the map would look like:

Answer:

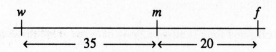

5.27   Three of the many recessive mutations in *Drosophila melanogaster* that affect body color, wing shape, or bristle morphology are black (*b*) body versus grey in the wild type, dumpy (*dp*), obliquely truncated wings versus long wings in the wild type, and hooked (*hk*) bristles at the tip versus not hooked in the wild type. From a cross of a dumpy female with a black and hooked male, all the F$_1$ were wild type for all three characters. The testcross of an F$_1$ female with a dumpy, black, hooked male gave the following results:

wild type	169
black	19
black, hooked	301
dumpy, hooked	21
hooked	8
hooked, dumpy, black	172
dumpy, black	6
dumpy	305
Total	1000

a.   Construct a genetic map of the linkage group (or groups) these genes occupy. If applicable, show the order and give the map distances between the genes.

b.   (1) Determine the coefficient of coincidence for the portion of the chromosome involved in the cross. (2) How much interference is there?

Answer:

a.   By doing a three-point mapping analysis as described in the chapter, we find that the order of genes in the chromosome is *dp-b-hk*, with 35.5 mu between *dp* and *b*, and 5.4 mu between *b* and *hk*.

b.   (1) The frequency of observed double crossovers is 1.4%, and the frequency of expected double crossovers is 1.9%. The coefficient of coincidence, therefore, is 1.4/1.9 = 0.73. (2) The interference value is given by 1 − coefficient of coincidence = 0.27.

5.28   In corn, colorless aleurone (*c*) is recessive to (*C*), shrunken (*sh*) is recessive to full (*Sh*), and waxy (*wx*) is recessive to starchy (*Wx*). The F$_1$ plants from the cross of true-breeding colored, shrunken, and starchy x true-breeding colorless, full, and waxy were crossed with colorless, shrunken, and waxy plants, and the following progeny were generated:

colored, shrunken, starchy	2538
colorless, full, waxy	2708
colored, full, waxy	116
colorless, shrunken, starchy	113
colored, shrunken, waxy	601
colorless, full, starchy	626
colored, full, starchy	4
colorless, shrunken, waxy	2

Map the positions of the *c*, *sh*, and *wx* genes in the chromosome.

Answer: This cross is a typical three-point-mapping cross. The parents were

$$\frac{C\ sh\ Wx}{C\ sh\ Wx} \qquad \text{and} \qquad \frac{c\ Sh\ wx}{c\ Sh\ wx}$$

The F$_1$ is
$$\frac{C\ sh\ Wx}{c\ Sh\ wx}$$

By inspecting the progeny from a cross of the F$_1$ with a triply homozygous recessive, we can find the double-crossover classes as the least frequent among the progeny; namely, the colored, full, starchy (*C Sh Wx*) and colorless, shrunken, waxy (*c sh wx*). Given the genotype of the F$_1$, the order of the genes is *c - sh* - wx. Thus the colored, full, waxy and colorless, shrunken, starchy classes derive from a single crossover between *c* and *sh*, and the colored, shrunken, waxy and colorless, full, starchy classes derive from a single crossover between *sh* and *wx*. Using the method of three-point mapping we have discussed in this chapter, the map distance between *c* and *sh* is 3.5 mu, and that between *sh* and *wx* is 18.4 mu. The resulting map is as shown:

5.29 In Chinese primroses long style (*l*) is recessive to short (*L*), red flower (*r*) is recessive to magenta (*R*) and red stigma (*rs*) is recessive to green (*Rs*). From a cross of homozygous short, magenta flower, and green stigma with long, red flower, and red stigma, the F$_1$ was crossed back to long, red flower, and red stigma. The following offspring were obtained:

Style	Flower	Stigma	Number
short	magenta	green	1063
long	red	red	1032
short	magenta	red	634
long	red	green	526
short	red	red	156
long	magenta	green	180
short	red	green	39
long	magenta	red	54

Map the genes involved.

Answer : The original cross was

$$\frac{L\ R\ Rs}{L\ R\ RS} \ \text{X} \ \frac{l\ r\ rs}{l\ r\ rs}$$

The testcross of the F$_1$ is a typical three-point-mapping cross. The total number of progeny is 3684. The double crossovers are the classes with 39 and 54 representatives. Along with the genotypes of the F$_1$, this result tells us that the order of genes is *l r - rs*. From the methods developed in the chapter the map is as shown:

5.30 The frequencies of gametes of different genotypes, determined by testcrossing a triple heterozygote, are as follows:

Gamete genotype			%
+	+	+	12.9
a	b	c	13.5
+	+	c	6.9
a	b	+	6.5
+	b	c	26.4
a	+	+	27.2
a	+	c	3.1
+	b	+	3.5
		Total	100.0

a. Which gametes are known to have been involved in double crossovers?
b. Which gamete types have not been involved in any exchanges?
c. The order shown is not necessarily correct. Which gene locus is in the middle?

Answer: a. $a + c$ and $+ b +$ (least frequent)
b. $+ b\ c$ and $a + +$ (most frequent)
c. locus c

5.31 Genes a, b, and c are recessive. Females heterozygous at these three loci are crossed to phenotypically wild-type males. The progeny are phenotypically as follows:

Daughters: all + + +

Sons:				
+	+	+		23
a	b	c		26
+	+	c		45
a	b	+		54
+	b	c		427
a	+	+		424
a	+	c		1
+	b	+		0
		Total		1000

a. What is known of the genotype of the females' parents with respect to these three loci? Give gene order and the arrangement in the homologs.
b. What is known of the genotype of the male parents?
c. Map the three genes.

Answer: a. $a^+ c\ b/a\ c^+ b^+$
b. $a^+ c^+ b^+/Y$
c.

$$a \qquad c \qquad\qquad b$$
$$\vdash\!\!-\!\!\!-\!\!\!-\!\!\dashv\!\!-\!\!\!-\!\!\!-\!\!\!-\!\!\dashv$$
$$\leftarrow 5 \rightarrow\!\leftarrow\!-\!-\ 10\ \longrightarrow$$

58

5.32 Two normal-looking Drosophila are crossed and yield the following phenotypes among the progeny:

Females:	+	+	+	2000

Males:	+	+	+	3
	a	b	c	1
	+	b	c	839
	a	+	+	825
	a	b	+	86
	+	+	c	90
	a	+	c	81
	+	b	+	75
			Total	4000

Give parental genotypes, gene arrangement in the female parent, map distances, and the coefficient of coincidence.

Solution: The disparity in results as a function of sex indicates that these genes are sex linked, i.e. located on the X chromosome. As a result, the phenotype of the male progeny reflects the genotype of the maternal X chromosome (any male offspring has the Y chromosome as its paternal contribution). Since the parents were "normal-looking", the male parent must have been + + + /Y. Since there are all possible classes of offspring that a heterozygote could produce, the female parent was heterozygous for all three loci. Inspection of the male progeny classes reveals that the two largest classes are + b c and a + +. Hence, the female parental genotype must have been + b c / a + +. Such a fly would be heterozygous at all loci and have a normal phenotype. Essentially the a and b loci are in repulsion; the b and c loci are in coupling.

The table is already arranged in an order that reflects reciprocal pairs, although the double crossovers are at the head of the list, followed by the non-recombinant parentals. Let us construct a table to show all cross-over frequencies, as follows:

				Recombination between			
				a and *b*	*b* and *c*	*a* and *c*	
+	b	c	839	0	0	0	parental
a	+	+	825	0	0	0	parental
a	b	+	86	86	86	0	
+	+	c	90	90	90	0	
a	+	c	81	0	81	81	
+	b	+	75	0	75	75	
+	+	+	3	3	0	3	
a	b	c	1	1	0	1	
Totals			2000	180	332	160	

We consider only male progeny because we can not score recombinants among the females. This is not a standard testcross. The frequency of recombination between a and b is 180/2000 = 9%. The frequency of recombination between b and c is 332/2000 = 16.6%. The frequency of recombination between a and c is 160/2000 = 8%. Since the frequency of recombination (map distance) is greatest between b and c they must mark the end points of the map; a is the central gene. Once again, this could have been deduced by inspection and comparison of the non-recombinant *vs.* double cross-over classes. A map of this section of X chromosome would look like:

```
 b              a          c
_|____9.0_____|___8.0__|_
```

To calculate the coefficient of coincidence we must compare the frequency of actual double crossovers to expected double crossovers. The actual frequency is (3 + 1)/2000 =

4/2000 = 0.002. The expected frequency of doubles is the product of the frequency of the singles: (0.09)(0.08) = 0.0072.

Answer: All three genes are X-linked. The parents are

$$\frac{b \quad + \quad c}{+ \quad a \quad +} \qquad \text{and} \qquad \frac{+ \quad + \quad +}{Y}$$

Map distances are $b - a = 9$ mu; $a - c = 8$ mu (progeny totals for linkage calculations is 2000 males). The coefficient of coincidence is 0.0020/0.0072 = 0.278.

5.33   Three different semidominant mutations affect the tail of mice. They are linked genes, and all three are lethal in the embryo when homozygous. Fused-tail (*Fu*) and kinky-tail (*Ki*) mice have kinky-appearing tails, while brachyury (*T*) mice have short tails. A fourth gene, histocompatibility-2 (*H-2*), is linked to the three tail genes and is concerned with tissue transplantation. Mice that are *H-2*/+ will accept tissue grafts, whereas +/+ mice will not. In the following crosses the normal allele is represented by a +. The phenotypes of the progeny are given for four crosses.

(1)	$\dfrac{Fu \quad +}{+ \quad Ki}$	x	$\dfrac{+ \quad +}{+ \quad +}$	Fused tail Kinky tail Normal tail Fused-kinky tail	106 92 1 1
(2)	$\dfrac{Fu \quad H\text{-}2}{+ \quad +}$	x	$\dfrac{+ \quad +}{+ \quad +}$	Fused tail, accepts grafts Normal rail, rejects graft Normal tail, accepts graft Fused tail, rejects graft	88 104 5 3
(3)	$\dfrac{T \quad H\text{-}2}{+ \quad +}$	x	$\dfrac{+ \quad +}{+ \quad +}$	Brachy tail, accepts graft Normal tail, rejects graft Brachy tail, rejects graft Normal tail, accepts graft	1048 1152 138 162
(4)	$\dfrac{Fu \quad +}{+ \quad T}$	x	$\dfrac{+ \quad +}{+ \quad +}$	Fused tail Brachy tail Normal tail Fused-brachy tail	146 130 14 10

Make a map of the four genes involved in these crosses, giving gene order and map distances between the genes.

Answer: The recombination percentages of the four crosses are calculated by taking into account the absence of certain classes owing to the recessive lethality. The percentages for the four crosses are (1) 1%, (2) 4%, (3) 12%, (4) 8%. The two possible maps to be drawn from these figures (it cannot be assumed automatically that the terminal markers are involved in a single cross, although as it turns out, they are) are as shown (next page):

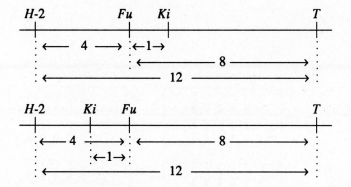

5.34 The cross in *Drosophila* of

$$\frac{a^+\ b^+\ c\ \ d\ \ e}{a\ \ b\ \ c^+\ d^+\ e^+} \times \frac{a\ \ b\ \ c\ \ d\ \ e}{a\ \ b\ \ c\ \ d\ \ e}$$

gave 1000 progeny of the following 16 phenotypes:

	Genotype	Number		Genotype	Number
(1)	$a^+\ b^+\ c\ \ d\ \ e$	220	(9)	$a\ \ b^+\ c^+\ d\ \ e^+$	14
(2)	$a^+\ b^+\ c\ \ d\ \ e^+$	230	(10)	$a\ \ b^+\ c^+\ d\ \ e$	13
(3)	$a\ \ b\ \ c^+\ d^+\ e$	210	(11)	$a^+\ b\ \ c\ \ d^+\ e^+$	8
(4)	$a\ \ b\ \ c^+\ d^+\ e^+$	215	(12)	$a^+\ b\ \ c\ \ d^+\ e$	8
(5)	$a\ \ b^+\ c^+\ d^+\ e$	12	(13)	$a^+\ b^+\ c^+\ d\ \ e^+$	7
(6)	$a\ \ b^+\ c^+d^+\ e^+$	13	(14)	$a^+\ b^+\ c^+\ d\ \ e$	7
(7)	$a^+\ b\ \ c\ \ d\ \ e^+$	16	(15)	$a\ \ b\ \ c\ \ d^+\ e^+$	6
(8)	$a^+\ b\ \ c\ \ d\ \ e$	14	(16)	$a\ \ b\ \ c\ \ d^+\ e$	7

a. Draw a genetic map of the chromosome, indicating the linkage of the five genes and the number of map units separating each.

b. From the single-crossover frequencies, what would be the expected frequency of $a^+\ b^+\ c^+\ d^+\ e^+$ flies?

Answer: a. *a,b,c*, and *d* are linked on the same chromosome, since the percentage of recombination is less than 50. *e* is on a separate chromosome, because it segregates independently from the four other genes. The map is constructed by calculating the recombination frequency for all possible pairs of *a,b,c,d*. The map distances allow an order to be determined. The map is shown below.

```
b               d         a          c              e
+---------------+---------+----------+              +
  <--- 5.5 --->  <- 4.3 ->  <- 2.7 ->
```

b. To get an $a^+\ b^+\ c^+\ d^+\ e^+$ fly, there must be a crossover between *b* and *d*, *d* and *a*, and *a* and *c*; 1/2 of the progeny from those crossovers are $a^+\ b^+\ c^+\ d^+$, and the other 1/2 are *a b c d*. Then 1/2 are e^+. The answer is 1.6×10^{-5}, i.e., 0.055 x 0.043 x 0.027 x 0.5 (for the half of the progeny produced by the triple crossover that have all the wild-type alleles) x 0.5 (to give the proportion that are e^+).

5.35 Complete all diagrams in the figure by correctly showing the centromeres, chromosome strands, and alleles on each strand.

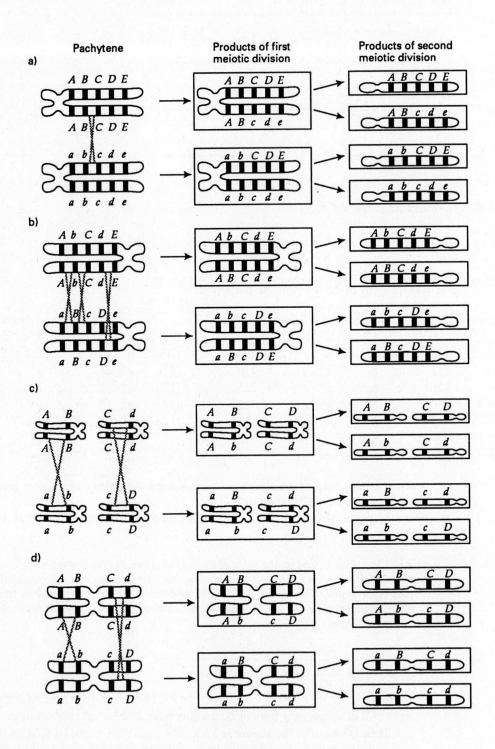

Answer:

CHAPTER 6

ADVANCED GENETIC MAPPING IN EUKARYOTES

I. CHAPTER OUTLINE

Tetrad analysis
    Life Cycle of Yeast
    Life Cycle of *Chlamydomonas reinhardi*
    Using Random–Spore Analysis to Map Genes in Haploid Eukaryotes
Using Tetrad Analysis to Map Two Linked Genes
Calculating Gene–Centromere Distance in Organisms With Linear Tetrads
Mitotic Recombination
    Discovery of Mitotic Recombination
    Mechanism of Mitotic Crossing–over
    Mitotic Recombination in Fungal Systems
Mapping Genes in Human Chromosomes
    Mapping Human Genes by Recombination Analysis
    Mapping Human Genes by Somatic Cell Hybridization

II. IMPORTANT TERMS AND CONCEPTS

tetrad analysis	heterokaryon
mating type	diploidization
ascus	haploidization
ascospore	parasexual system
parental ditype (PD)	somatic cell hybridization
tetratype (T)	HAT technique
nonparental ditype (NPD)	synteric
first division segregation	synkaryon
second division segregation	chromosome deletion
mitotic crossing-over	chromosome translocation
mosaic phenotype	

III. THINKING ANALYTICALLY

When dealing with problems in tetrad analysis, it is a good idea at the start: (1) to diagram the tetrads, as shown below. Darken one of the centromeres of the pair in order to avoid making meaningless crossovers between sister chromatids. (Why would they be meaningless?)

    (2) For solving Text Problem 6.1, for example, assume genes *a* and *b* and their respective wild type alleles, + and +. Until you get used to using + for the wild type allele for any gene, add a superscript so you won't confuse which is the + for *a*, and which is the + for *b*, like this: $+^a$, $+^b$. Now, since the "meiotic event" that leads to the three recombination patterns designated PD, NPD, and T is crossing-over, diagram each of these types, as shown in the following figure:

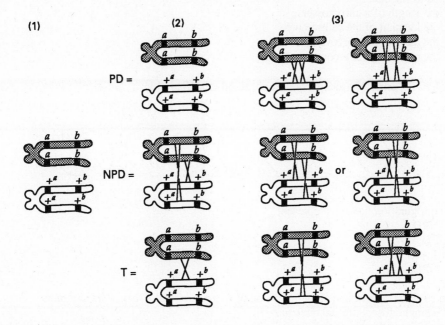

(3) Then, starting at the centromere end, trace along each chromatid, crossing over where you need to in order to achieve each of the patterns required. After making a crossover, resume tracing along in the same direction - away from the centromere end - until you come to the end of the chromatid.

The answer to the text question, "..what meiotic events give rise to PD, NPD, and T tetrads?" therefore is this: Noncrossovers and two-strand double crossovers produce PDs; four-strand double crossovers produce NPDs; and single crossovers and three-strand double crossovers yield Tetratypes.

IV. QUESTIONS FOR PRACTICE

A. MULTIPLE CHOICE QUESTIONS

1. Tetrad analysis is a technique for
   a. determining base sequences in DNA.
   b. measuring the number of chromatids at meiosis.
   c. separating *Neurospora* mycelia.
   d. mapping genes in certain organisms.

2. Ordered tetrads occur in the asci of
   a. *Neurospora.*
   b. *Chlamydomonas.*
   c. *Saccharomyces.*
   d. *Drosophila.*

3. A single cross-over between two gene loci during meiosis in *Neurospora* spore formation produces in the resulting ascospores a genetic configuration known as a
   a. tetratype.
   b. parental ditype.
   c. nonparental ditype.
   d. double recombinant.

4. The formula $\dfrac{1/2 \ (T) + NPD}{\text{total \# tetrads}} \times 100$

   is used to calculate:
   a. the length of a chromosome.
   b. the location of the centromere.
   c. the number of crossovers.
   d. the percentage of recombinants.

5. The following array of ascospores was observed in a *Neurospora* ascus following the cross *a* + x + *b*:

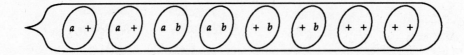

   What configuration does this represent?
   a. NPD.
   b. PD.
   c. T.
   d. NPD + PD.

6. A heterokaryon cell produced by the fusion of two mycelia of *Aspergillus* contains
   a. a single, fused, diploid nucleus.
   b. two fused diploid nuclei.
   c. two unfused haploid nuclei.
   d. two fused haploid nuclei.

7. The HAT-technique
   a. causes selected human and mouse cells to fuse.
   b. segregates hybridized human/mouse cells from unfused human and mouse cells.
   c. causes certain human chromosomes to be eliminated from hybridized cells.
   d. identifies the genotypes of selected human chromosomes.

Answers: 1d, 2a, 3a, 4d, 5c, 6c, 7b

B. Thought Question

1. Tests for the presence of the gene for Human Clotting Factor 7 (HCF-7) in certain somatic cell hybrid lines gave the results shown in the table below. A + or − indicates the presence or absence of a chromosome in each cell line, and the presence or absence of the clotting factor. In which chromosome is the gene located?

Cell Line	Chromosome Number								HCF-7
	1	2	4	6	8	14	15	20	
1	+	−	−	+	+	−	+	−	+
2	+	−	+	+	−	−	−	+	−
3	−	+	-	+	+	+	−	+	+
4	+	−	−	+	+	−	+	+	+
5	−	+	+	+	−	−	−	+	−

Answer: Chromosome 8.

V. ANSWERS AND SOLUTIONS TO TEXT PROBLEMS

6.1 In *Saccharomyces*, *Neurospora*, and *Chlamydomonas*, what meiotic events give rise to PD, NPD, and T teterads?

Answer: PD: noncrossover and two-strand double crossovers; NPD: four-strand double crossovers; T: single crossovers and three-strand double crossovers.

6.2 What important item of information regarding crossing-over can be obtained from tetrad analysis (as in *Neurospora*) but not from single-strand analysis (as in *Drosophila*)?

Answer: That crossing-over occurs at the four-strand stage of meiosis.

6.3 A cross was made between a pantothenate-requiring (*pan*) strain and a lysine-requiring (*lys*) strain of *Neurospora crassa*, and 750 random ascospores were analyzed for their ability to grow on a minimal medium (a medium lacking panthothenate and lysine). Thirty colonies subsequently grew. Map the *pan* and *lys* loci.

Answer: The 30 colonies represent half the recombinants since the double mutants cannot grow on the minimal medium. The map distance between *pan* and *lys* is given by (60/750) x 100% = 8% (8 mu).

6.4 In *Neurospora* the following crosses yielded the progeny as shown:

$$a^+ b \text{ x } a b^+ \rightarrow$$

981	a^+	b
1000	a	b^+
10	a^+	b^+
9	a	b
2000		

$$a^+ c \text{ x } a c^+ \rightarrow$$

850	a^+	c
833	a	c^+
169	a^+	c^+
148	a	c
2000		

$$b^+ c \text{ x } b c^+ \rightarrow$$

850	b^+	c
850	b	c^+
140	b^+	c^+
160	b	c
2000		

What is the probable gene order and what are the approximate map distances betwen adjacent genes?

Answer:

6.5　Four different albino strains of *Neurospora* were each crossed to the wild type. All crosses resulted in half wild-type and half albino progeny. Crosses were made between the first strain and the other three with the following results:

$$
\begin{array}{lll}
1 \times 2: & 975 \text{ albino, } 25 \text{ wild} \\
1 \times 3: & 1000 \text{ albino} \\
1 \times 4: & 750 \text{ albino, } 250 \text{ wild type}
\end{array}
$$

Which mutations represent different genes, and which genes are linked? How did you arrive at your conclusions?

Answer:　Genes 1 and 2 are linked, since they yield much less than 25% wild type; 25% wild type is expected on the basis of independent assortment, and other recombinants - the double mutants - being among the albino progeny. The 25 individuals that are wild type in the 1 x 2 cross are presumably accompanied by the albino, reciprocal class (double mutants) in approximately equal frequency. The true recombination frequency is thus about 50/1000, or 5%. The expectation of independent assortment is realized in the 1 x 4 cross. The cross 1 x 3 indicates, as far as one can tell, that the two mutations are allelic or at least very closely linked; they fail to produce any recombinants among a sizable number of progeny.

6.6　Genes *met* and *thi* are linked in *Neurospora crassa*. We wish to locate *arg* with respect to *met* and *thi*. From the cross *arg* + + x + *thi met*, the following random ascospore isolates were obtained. Map these three genes.

arg	*thi*	*met*	26		*arg*	+	+	51
arg	*thi*	+	17		+	*thi*	+	4
arg	+	*met*	3		+	+	*met*	14
+	*thi*	*met*	56		+	+	+	29

Answer:

```
arg                        met        thi
 +--------------------------+----------+
      <--------- 31 -------> <--- 19 --->
```

6.7　Given a *Neurospora* zygote of the constitution shown in the figure below, diagram the significant events producing an ascus where the *A* alleles segregate at the first division and the *B* alleles segregate at the second division.

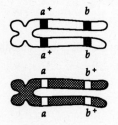

68

Answer: The text figure can be represented more easily, albeit less realistically, by the shorthand diagram at the left. The solution is in the middle, and its resolution at the right:

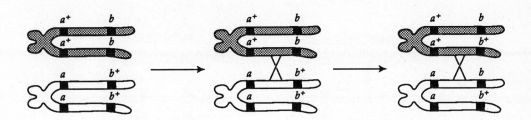

6.8   Double exchanges between two loci can be of several types, called two-strand, three-strand, and four-strand doubles.

    a.   Four recombination gametes would be produced from a tetrad in which the first of two exchanges was as depicted in the first figure below. Draw in the second exchange.

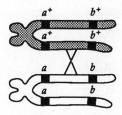

    b.   In the following figure, draw in the second exchange so that four nonrecombination gametes would result.

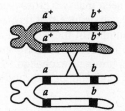

    c.   Other possible double-crossover types would result in two recombination and two parental gametes. If all possible multiple-crossover types occur at random, the frequency of recombination between genes at two loci will never exceed a certain percentage, regardless of how far apart they are on the chromosome. What is that percentage? Explain the reason why the percentage cannot exceed that value.

Solution: a) Draw the tetrads as below, and put in crosses that show where the crossovers are taking place. Let the cross at the right represent the crossover position given, and the one to the left the site of the second crossover that will give the desired result. Then to the right of this tetrad, draw the tetrad as it will appear after the exchange has taken place.

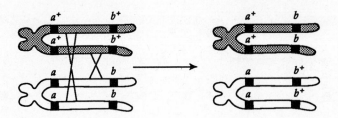

(b)   Repeat this procedure, but you will see that in order to effect no recombination, the second crossover must nullify the first - that is, you will need a double-crossover between the same two chromatids, as follows:

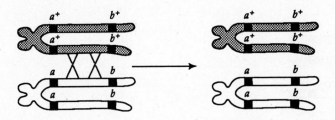

(c)   A third variety of a double exchange between the two loci a and b would be to make the second crossover between the two bottom strands, and a fourth variety, which would amount to the same thing, would be to make it between the two upper strands. If you do these, you will see that in each case, you will get two recombinants and two nonrecombinants, which net 50% recombination. From solutions to parts (a) and (b), above, you will notice that the net recombination here too is 50%. You have already seen that two crossovers between the same two strands net zero recombination. From this you can deduce that however many crossovers you have, and in whatever combination, any odd-number of these combinations will yield the results already achieved - that is, a maximum of 50% recombination - and any even-number will nullify whatever exchange was of effected by a previous exchange. Therefore the answer to (c), is 50% maximum, which is the equivalent of independent assortment.

6.9   A cross between a pink (p^-) yeast strain of mating type mt^+ and a grey strain (p^+) of mating type mt^- produced the following tetrads:

Number of Trends	Kind of Tetrad
18	$p^+ \, mt^+, p^+ \, mt^+, p^- \, mt^+, p^- \, mt^-$
8	$p^+ \, mt^+, p^- \, mt^+, p^+ \, mt^-, p^- \, mt^-$
20	$p^+ \, mt^-, p^+ \, mt^-, p^- \, mt^+, p^- \, mt^+$

On the basis of these results, are the p and mt genes on separate chromosomes?

Answer: The two genes can be considered to assort independently since the number of parental ditype (20) is approximately equal to the number of nonparental ditype (18).

6.10 In *Neurospora* the peach gene (*pe*) is on one chromosome and the colonial gene (*col*) is on another. Disregarding the occurrence of chiasmata, what kinds of tetrads (asci) would you expect and in what proportions if these two strains are crossed?

Solution: This problem assumes that one strain is $pe^+ col^-$ and the other is $pe^- col^+$. When crossed, a heterokaryon (diploid) cell is formed. Replication of chromatids occurs, followed by synapsis of homologues and alignment on the division spindle. Two different alignments are possible - with the pe^- and col^+ chromatids on one side of the metaphase plate, and pe^+ and col^- on the other, or with pe^- and col^- opposite pe^+ and col^+. In the first case, after the two divisions, the ascus would be: $pe^- col^+, pe^- col^+, pe^- col^+, pe^- col^+, pe^+ col^-, pe^+ col^-, pe^+ col^-, pe^+ col^-$. In the second case, the ascus would be: $pe^- col^-, pe^- col^-, pe^- col^-, pe^- col^- pe^+ col^+, pe^+ col^+, pe^+ col^+, pe^+ col^+$.

Answer: The two types of tetrads (asci) thus shown would be expected to occur with equal frequency.

6.11 The following unordered asci were obtained from the cross *leu* + x + *rib* in yeast. Draw the linkage map and determine the map distance.

110	45	6	39
leu +	*leu* *rib*	+ +	*leu* +
+ *rib*	*leu* +	*leu* *rib*	+ *rib*
leu +	+ +	*leu* *rib*	+ +
+ *rib*	+ *rib*	+ +	*leu* *rib*

Solution:
PD: (*leu* + ; + *rib*) = 110
NPD: (*leu* *rib* ; + +) = 6
T: (*leu* +; *leu* *rib*; + + ; + *rib*) = 45 + 39 = 84
Total: 110 + 6 + 84 = 200

$$\% \text{ recombination} = \frac{(1/2)(T) + NPD}{\text{Total \# Asci}} \times 100\%$$

$$= \frac{(1/2)(84) + 6}{200} \times 100\%$$

$$= 24\%$$

Answer: Map distance between *leu* and *rib* = 24 mu

6.12 The genes *a*, *b*, and *c* are on the same chromosome arm in *Neurospora crassa*. The following ordered asci were obtained from the cross *a b* + x + + *c*

45	5	146	1
a b c	*a b* +	*a b* +	*a b* +
+ *b c*	*a* + +	*a b* +	+ + +
a + +	+ *b c*	+ + *c*	*a b c*
+ + *c*	+ + *c*	+ + *c*	+ + *c*

10	20	15	58
a b +	*a b* +	*a b* +	*a b* +
a + *c*	+ + *c*	*a b c*	+ *b* +
+ *b* +	*a b* +	+ + +	*a* + *c*
+ + *c*	+ + *c*	+ + *c*	+ + *c*

a. Determine the correct gene order.

71

a. Determine the correct gene order.
b. Determine all gene-gene and gene-centromere distances.

Answer: a) From applying the tetrad analysis formula,

$$\text{map distance} = \frac{(1/2)\ (T) + NPD}{\text{Total \# Asci}} \times 100\%$$

the $a-b$ distance is 19.6 map units, the $b-c$ distance is 11 map units, and the $a-c$ distance is 14 map units. Thus, the gene order is $a-c-b$.

b) Centromere distances are calculated from the formula (% second division segregation tetrads) ÷ 2. Since we have all gene-gene distances, all we need to know is whether a or b is nearer the centromere. a is 20.7 map units from the centromere and b is 6 map units. Thus the map is as shown:

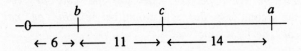

6.13   If 10 percent of the asci analyzed in a particular two-point cross of *Neurospora crassa* show that an exchange has occurred between two loci (i.e.. exhibit second division segregation), what is the map distance between the two loci?

Answer: Map distance is 1/2 x % of second division segregation asci which in this case equals 5 map units.

6.14   A diploid strain of *Aspergillus nidulans* (forced between wild type and a multiple mutant) that was heterozygous for the recessive mutations y (yellow), w (white), ad (adenine), sm (small), phe (phenylalanine), and pu (putrescine) produced haploid segregants. Forty-one haploid white and yellow segregants were tested and found to have the following genotypes and numbers:

| white: | y | w | pu | ad | sm | phe | 7 |
| | y | w | pu | ad | + | + | 11 |

| yellow: | y | + | + | + | sm | phe | 16 |
| | y | + | + | + | + | + | 7 |

What are the linkage relationships of these genes?

Answer: Haploidization gives an indication of linkage groups, as discussed in the text. The data indicate that there are three separate linkage groups involved, since they appear in haploids as independent "blocks" of genes: y, w-pu-ad, and sm-phe.

6.15   A heterokaryon was established in the fungus *Aspergillus nidulans* between a mer⁻, trp⁻ auxtroph and a leu⁻, nic⁻ auxtroph. A diploid strain was selected from this heterokaryon. From this diploid strain, the following eight haploid strains were obtained from conidial isolates, via the parasexual cycle, in approximately equal frequencies:

1.	nic^+ leu^+ met^- trp^-	5.	leu^+ nic^- met^+ trp^+
2.	met^+ leu^+ nic^+ trp^+	6.	met^- nic^- leu^- trp^-
3.	trp^- met^- leu^+ nic^-	7.	trp^+ leu^- met^+ nic^+
4.	leu^- nic^- trp^+ met^+	8.	nic^+ met^- trp^- leu^-

Which, if any, of these four marker genes are linked, and which are unlinked?

Answer: In haploidization, those genes that are linked always remain together in the segregants, while genes that are in a different linkage group segregate independently. No crossing-over is involved in haploidization. By inspection of the data, the *met* and *trp* genes are linked (i.e., all the segregants are either *met⁻ trp⁻* or *met⁺ trp⁺*) while the *nic* and *leu* genes segregate independently of the *met trp* linkage group and of each other. Therefore, there are three linkage groups involved: one with *nic*, one with *leu*, and the third with *met* and *trp*.

6.16 A (green) diploid of *Aspergillus nidulans* is heterozygous for *each* of the following recessive mutant genes: *sm, pu, phe, bi, w* (white), *y* (yellow), and *ad*. Analysis of white and yellow haploid segregants from this diploid indicated several classes with the following genotypes:

Genotype

sm	pu	phe	bi	w	y	ad
sm	pu	phe	+	w	y	ad
+	pu	+	+	w	y	ad
+	pu	+	bi	w	+	ad
sm	+	phe	+	+	y	+
+	+	+	+	+	y	+
sm	pu	phe	bi	w	+	ad

How many linkage groups are involved, and which genes are on which linkage group?

Answer: There are three linkage groups, as determined by which blocks of genes always stay together during the haploidization process: *sm-phe*, *pu-w-ad*, and *y-bi*.

6.17 A (green) diploid of *Aspergillus nidulans* is homozygous for the recessive mutant gene *ad* and heterozygous for the following recessive mutant genes: *paba, ribo, y* (yellow), *an, bi, pro,* and *su-ad*. Those recessive alleles that are on the same chromosome are in coupling. The *su-ad* allele is a recessive suppressor of the *ad* allele: The *+/su-ad* genotype does not suppress the adenine requirement of the *ad/ad* diploid, whereas the *su-ad/su-ad* genotype does suppress that requirement. Therefore the parental diploid requires adenine for growth. From this diploid two classes of segregants were selected: yellow and adenine-independent. The accompanying table lists the types of segregants obtained.

Segregant Type Selected	Phenotype					
Adenine-independent	+					
	ribo					
	ribo an					
	ribo an pro	paba	y		bi	
yellow			y	ad	bi	
		paba	y	ad	bi	
	pro	paba	y	ad	bi	
	ribo an pro	paba	y		bi	

Analyze these results as completely as possible to determine the location of the centromere and the relative locations of the genes.

Answer: The last class in each segregant type (*ribo, an, pro, paba, y, bi* in each case) represents haploid segregants and indicates that all of the genes under investigation are in the same chromosome. The rest of the classes are diploid segregants representing various mitotic recombinants.

Consider the adenine-independent segregants: Following the logic developed in the chapter for ordering genes in a chromosome arm - that is, all genes distal to a crossover become homozygous - we can deduce that the genes *an* and *ribo* are on the same chromosome arm as *su-ad* and that the order of genes is centromere–*an*–*ribo*–*su-ad*.

Similarly, analysis of the yellow diploid segregants shows that the order of genes in that chromosome arm is centromere–*pro–paba–y–(ad–bi)*. Since all yellow segregants are *ad* and *bi*, the latter two loci must be distal to *y*, but their relative order cannot be determined from the data given.

6.18   The accompanying table shows the only human chromosomes present in stable human-mouse cell hybrid lines.

		Human Chromosomes		
		2	4	10	19
	A	–	+	+	–
Hybrid	B	+	–	+	+
lines	C	–	+	+	+
	D	+	+	–	–

The presence of four enzymes, I, II, III, and IV, was investigated: I was present in *A*, *B*, and *C* but absent in *D*; II was in *B* and *D* but absent in *A* and *C*; III was in *A*, *C*, and *D* but no in *B*; and IV was in *B* and *C* but not in *A* and *D*. On what chromosomes are the genes for the four enzymes?

Solution: Look for a pattern of +'s in each of the vertical columns that corresponds with the rows designated for each of the enzymes. Thus enzyme (gene) I shows the pattern under chromosome 10 (+ + + –), II shows the pattern – + which is under chromosome 2 and therefore is located on chromosome 2. In this way you will find that III is on chromosome 4, and IV is on number 19.

Answer: I is on 10; II is on 2; Ill is on 4; IV is on 19.

CHAPTER 7

GENETIC RECOMBINATION IN BACTERIA AND BACTERIOPHAGES

I. CHAPTER OUTLINE

II. IMPORTANT TERMS AND CONCEPTS

colony	*Hfr* strain
transformation	*F'* (*F* prime)
transformant	*F* duction
natural *vs.* engineered	exconjugant
transformation	bacteriophage (phage)
cotransformation	phage lysate
conjugation	lytic cycle
transconjugant	virulent phage
transduction	lysogeny
transductant	lysogenic
auxotroph	lysogenic pathway
prototroph	prophage
plasmid	temperate phage
F factor	phage vector
F factor origin	cotransduction
F pili (sex pili)	plaque

III. THINKING ANALYTICALLY

With this chapter we undertake some rather complicated material that in places is a bit difficult to grasp intuitively. For example, there are many refinements of phage types and special characteristics. Go over the chapter sections that describe and name them. The names are quite descriptive as you will find out when you use them a few times when solving problems. It is not just a matter of rote memorization, however - you must use the terms in their context until you fix what is practically a visual image of them as they do their "thing".

As with preceding chapters, read the statement and conditions of the problem so that you have a clear understanding of it. Then organize the data carefully on paper, using a pencil, preferably, because you may have to reshuffle some or even all of it.

Go over the sample questions and their solutions in the section Analytical Approaches for Solving Genetics Problems, which appears at the end of the chapter. It will pay off to do so. They are very good.

IV. QUESTIONS FOR PRACTICE

A. Multiple Choice Questions

1. A strain, such as a strain of *E. coli*, that requires nutritional or other kind of supplements for growth and/or survival is known as
    a. a prototroph.
    b. a heterotroph.
    c. a pleiotroph.
    d. an auxotroph.

2. Bacterial cells that transfer both *F* factor and chromosomal genes during conjugation is identified as
    a. F^+.
    b. F'.
    c. *Hfr*.
    d. F^-.

3. The process by which cells take up genetic material from the extracellular milieu and incorporate it into their genetic complement is called
    a. transduction.
    b. transformation.
    c. translocation.
    d. DNA transfusion.

4 . Conjugation in bacteria involves
    a. the union of bacterial gametes.
    b. the fusion of two cells of opposite mating types.
    c. a virus mediated exchange of DNA.
    d. the transfer of DNA from one cell to another.

5. A self -replicating genetic element found in the cytoplasm of bacteria is a
    a. sex-pilus.
    b. plasmid.
    c. viral capsid.
    d. cotransductant.

6. The *F* factor possesses a region of DNA that contains fertility genes and is required for its self-replication within a host cell. This region is the
    a. *F* factor origin.
    b. *F* terminus.
    c. initial base pair.
    d. insertion element.

7. The formation of *F*-pili by conjugating bacteria is specified by genes
   a. located in the donor cell's cytoplasm.
   b. activated by sex hormones.
   c. in the donor bacterium's chromosome.
   d. present in the *F* factor.

8. A characteristic of *Hfr* strains is that they rarely convert an *F⁻* cell to *Hfr* because
   a. they require a viral vector.
   b. the necessary genes for their *F* factor are located partly at the beginning and partly at the end of the donor chromosome.
   c. the donor chromosome is rarely transferred completely during conjugation.
   d. both choices, b and c, above, apply

9. Prophage is a virus particle that is
   a. integrated into the host's chromosome.
   b. replicating in the host cell's cytoplasm.
   c. in the process of infecting a host cell.
   d. virulent.

10. Phages that can be in either a lytic or lysogenic pathway are called
    a. virulent
    b. temperate
    c. intemperate
    d. transductant

Answers: 1d, 2b, 3b, 4d, 5b, 6a, 7d, 8d, 9a, 10b

B. Thought Questions

1. Explain why two genes that are close together on a chromosome show a high frequency of cotransduction. (cf. text p. 211)
2. Compare the life cycles of T4 and λ. (cf. Figures 7.14, 7.15)
3. Contrast the features and/or conditions of generalized transduction and specialized transduction. (cF.text pp 208-214)
4. Comment on the significance of bacterial transformation and conjugation in relation to the rapidity with which certain infectious organisms adapt to changing environmental conditions or factors, such as antibiotics. (Hint: Sexual reproduction promotes genetic variability. How do parasexual systems compare?)

V. ANSWERS AND SOLUTIONS TO TEXT QUESTIONS

7.1 If an *E. coli* auxotroph *A* could only grow on a medium containing thymine, and an auxotroph *B* could only grow on a medium containing leucine, how would you test whether DNA from *A* could transform *B*?

Answer: Add DNA extract from *A* to a leucine-fortified culture of *B*. Incubate long enough for transformation to occur, then plate out on minimum medium. Any colonies that appear will consist of transformed organisms wild-type at both loci.

7.2 Distinguish among *F⁻*, *F⁺*, *F'*, and *Hfr* strains of *E. coli*.

Answer: See *F⁻* (p. 200); *F⁺* (p. 200); *F'* (p. 202); *Hfr* (p. 202).

7.3　In F^+ x F^- crosses the F^- recipient is converted to a donor with very high frequency. However, it is rare for a recipient to become a donor in Hfr x F^- crosses. Explain why.

Answer:　The whole chromosome would have to be transferred in order for the recipient to become a donor in an Hfr x F^- cross; that is, the F factor in the Hfr strain is transferred to the F^- cell last. This transfer takes approximately 100 min, and usually, the conjugal unions break apart before then.

7.4　With the technique of interrupted mating four Hfr strains were tested for the sequence in which they transmitted a number of different genes to an F^- strain. Each Hfr strain was found to transmit its genes in a unique sequence, as shown in the accompanying table (only the first six genes transmitted were scored for each strain).

Order of Transmission		Hfr Strain		
	1	2	3	4
First	O	R	E	O
	F	H	M	G
	B	M	H	X
	A	E	R	C
	E	A	C	R
Last	M	B	X	H

What is the gene sequence in the original strain from which these Hfr strains derive? Indicate on your diagram the origin and polarity of each of the four Hfr's.

Answer:

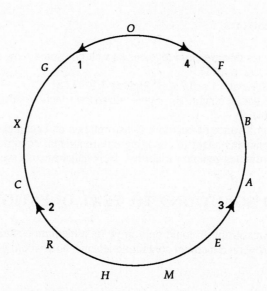

7.5　When Hfr donors conjugate with F^- recipients that are lysogenic for phage λ, the recipients (the exconjugant zygotes) usually survive. However, when Hfr donors that are lysogenic for λ conjugated with F^- cells that are nonlysogens, the exconjugant zygotes produced from matings that have lasted at least 100 minutes usually lyse, releasing mature phage particles. This event is called zygotic induction of λ.
　　a.　Explain zygotic induction.
　　b.　Explain how the locus of the integrated λ prophage can be determined.

Answer: a. When the prophage enters the enters the recipient (F^-) there are no repressor molecules present so the prophage is induced, it goes through the lytic cycle, and progeny phage are released. It is called zygotic induction since the initiation of the lytic cycle depends on the formation of a zygote (conjugal union) between the donor and recipient. As long as the prophage (i.e., the λ gene set) is still within the donor, there are enough repressor molecules present to keep the genome in the prophage state.

       b. The λ genome can be mapped just like any other gene by determining the time at which it enters the recipient. As we have seen, as soon as it enters the recipient, the lytic cycle is induced and progeny phages are released from the lysed cell. The way to map the prophage site, then, is to do an interrupted-mating experiment and, at various sample times, plate exconjugants on suitable plates that will select for those F^- that have received donor markers, that will select against both parentals, and that contain a lawn of sensitive bacteria on which plaques of λ can be detected. Up to a certain time no plaques will be seen; then once the λ is transferred, plaques will be seen.

7.6   At time zero an *Hfr* strain (strain 1) was mixed with an F^- strain, and at various times after mixing, samples were removed and agitated to separate conjugating cells. The cross may be written as:

$$Hfr\ 1: \quad a^+\ b^+\ c^+\ d^+\ e^+\ f^+\ g^+\ h^+\ str^s$$
$$F^-: \quad a^-\ b^-\ c^-\ d^-\ e^-\ f^-\ g^-\ h^-\ str^r$$

(No order is implied in listing the markers.)

The samples were then plated onto selective media to measure the frequency of $h^+\ str^r$ recombinants which had received certain genes from the *Hfr* cell. The graph of the number of recombinants against time is shown in the following figure:

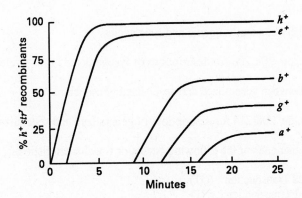

79

a. Indicate whether each of the following statements is true or false.

i. All F^- cells which received a^+ from the Hfr in the chromosome transfer process must also have received b^+.

ii. The order of gene transfer from Hfr to F^- was a^+ (first), then g^+, then b^+, then e^+, then h^+.

iii. Most e^+ str^r recombinants are likely to be Hfr cells.

iv. None of the b^+ str^r recombinants plated at 15 minutes are also a^+.

b. Draw a linear map of the Hfr chromosome indicating the
   i.   point of insertion, or origin;
   ii. the order of the genes a^+, b^+, e^+, g^+, and h^+;
   iii. the shortest distance between consecutive genes on the chromosomes.

Answer:

a. i. True. The graph shows that the transfer of the chromosome from the Hfr donor has been linear and sequential, starting at the h^+ locus and proceeding at least as far as a^+. It follows that all loci between h^+ and a^+ are likely to have been transferred, including b^+.

ii. False. The data show that a^+ took the longest time to be transferred and so was the last gene to be received.

iii. False. In order for the recipient cell to be transformed to F^+ or Hfr, it must receive a complete copy of the F factor. Actually only part of the F factor is transferred at the beginning of conjugation, the rest being at the end of the donor chromosome. Because of turbulence, the chromosome almost invariably is broken before transfer can be completed, and so the probability that complete transfer will occur is quite remote. Experimental data bear this out.

iv. True. The graph shows that a^+ is transferred after about 16 or 17 minutes elapsed time.

b. Time:

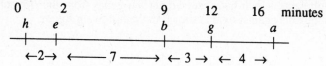

7.7 Distinguish between the lysogenic and lytic cycles.

Answer: See text, pp. 206-208 for descriptions of lysogenic and lytic cycles.

7.8 Distinguish between generalized and specialized transduction.

Answer: See text, pp. 208-214 for descriptions of generalized and specialized transduction.

7.9 Indicate whether each of the following occurs/or is a characteristic of

Generalized Transduction (GT)
Specialized Transduction (ST)
Occurs in Both (B)
Occurs in Neither (N)

a. Phage carries DNA of bacterial or viral DNA origin, never both.
b. Phage carries viral DNA covalently linked to bacterial DNA.
c. Phage integrates into a specific site on the host chromosome.
d. Phage integrates at a random site on the host chromosome.
e. "Headful" of bacterial DNA encoated.
f. Host lysogenized.
g. Prophage state exitsts.
h. Temperate phage involved.
i. Virulent phage involved.

a. GT
b. ST
c. ST
d. GT
e. GT
f. ST
g. ST
h. B
i. N

7.10   Consider the following transduction data:

Donor		Recipient		Selected Marker	Unselected Marker	%
$aceF^+$	dhl	aceF	dhl^+	$aceF^+$	dhl	88
$aceF^+$	leu	aceF	leu^+	$aceF^+$	leu	34

Is *dhl* or *leu* closer to *aceF*?

Answer: The notation "*aceF*" represents a specific insertion site for an *F* factor. This table shows that cells selected for transduction of F^+ were isolated and then tested for *dhl* and *leu*. It shows that 88% of the $aceF^+$ cells isolated included the *dhl* marker, while 34% included *leu*. Recall that the closer two loci are together on the same chromosome, the greater is the probability that they will both be included within the limits of a double crossover. Recall also that a double crossover is a necessary condition for recombination. Therefore, *dhl* is closer to the $aceF^+$ marker than *leu*.

7.11   Consider the following P1 transduction data:

Donor		Recipient		Selected Marker	Unselected Marker	%
$cysB^+$	trpE	cysB	$trpE^+$	$cysB^+$	trpE	37
$cysB^+$	trpB	cysB	$trpB^+$	$cysB^+$	trpB	53

Is *trpE* or *trpB* closer to *cysB*?

Answer: The *trpB* marker cotransduced with the $cysB^+$ selected marker more frequently than did the *trpE*; hence *trpB* is closer to *cysB*.

7.12   Consider the following data with P1 transduction:

Donor		Recipient		Selected Marker	Unselected Marker	%
aroA	$pyrD^+$	$aroA^+$	pyrD	$pyrD^+$	aroA	5
$aroA^+$	cmlb	aroA	$cmlB^+$	$aroA^+$	cmlB	26
cmlB	$pyrD^+$	$cmlB^+$	pyrD	$pyrD^+$	cmlB	54

Choose the correct order:

a.   *aroA–cmlB–pyrD*
b.   *aroA –pyrD–cmlB*
c.   *cmlB–aroA–pyrD*

81

Answer: (a) shows the correct order, i.e., *aroA–cmlB–pyrD*. The *pyrD* gene is quite far from the *aroA* gene since only 5% cotransduction occurred. The *cmlB* gene cotransduced with both *aroA* (26%) and *pyrD* (54%). Taken together, the data place the *cmlB* gene between the other two genes.

7.13 Order the mutants *trp*, *pyrF* and *qts* on the basis of the following three factor transduction cross:

Donor	*trp*$^+$	*pyr*$^+$	*qts*
Recipient*trp*	*pyr*	*qts*$^+$	
Selected Marker	*trp*$^+$		

Unselected Markers		Number
pyr$^+$	*qts*$^+$	22
pyr$^+$	*qts*	10
pyr	*qts*$^+$	68
pyr	*qts*	0

Answer: The principle of assigning gene order is to relate the number of crossovers that would be necessary to produce the recombinants to the frequency of recombinants. Recombinants produced by a double crossover event (the minimum needed to produce a transductant) would be expected to occur at a much higher frequency than recombinants produced by a quadruple crossover event. In this case we can draw a gene order that requires a quadruple crossover to produce the *trp*$^+$ *pyr qts* class and that is *trp pyr qts*. The other three transductant classes can be generated by double crossovers.

7.14 Order *cheA*, *cheB*, *eda* and *supD* from the following data:

Markers	% cotransduction
cheA–eda	15
cheA–supD	5
cheB–eda	28
cheB–supD	2.7
eda–supD	0

Answer: Working from the principle that a relatively high cotransduction frequency means that the genes are relatively close, the order of the genes is: *supD–cheA–cheB–eda*.

7.15 A stock of T4 phage is diluted by a factor of 10^{-8} and 0.1 mL of it is mixed with 0.1 mL of 10^8 *E. coli B*/mL and 2.5 mL melted agar, and poured on the surface of an agar petri dish. The next day 20 plaques are visible. What is the concentration of T4 phages in the original T4 stock?

Answer: Only the volume of phage suspension plated is relevant to the calculation. 0.1 mL of the diluted phage suspension produced 20 plaques, meaning there were 200 phages/mL of the diluted suspension. The dilution factor was 10^8, so the concentration in the original suspension was 200 x 10^8 or 2 x 10^{10}.

7.16 Wild-type phage T4 grows on both *E. coli B* and *E. coli K12*(λ), producing turbid plaques. The *rII* mutants of T4 grow on *E. coli B*, producing clear plaques, but do not grow on *E. coli K12*(λ). This host range property permits the detection of a very low number of r^+ phages among a large number of *rII* phages. With this sensitive system it is possible to determine the genetic distance between two mutations within the same gene, in this case the *rII* locus. Suppose *E. coli B* is mixedly infected with *rIIx* and *rIIy*, two separate mutants in the *rII* locus. Suitable dilutions of progeny phages are plated on *E. coli B* and *E. coli K12*(λ). A 0.1-mL sample of a thousandfold dilution plated on *E. coli B* showed 672 plaques. A 0.2-mL sample of undiluted phage plated on *E. coli K12*(λ) showed 470 turbid plaques. What is the genetic distance between the two *rII* mutations?

Answer: 0.07 mu. The plaques produced on *K12* (λ) are wild-type r^+ phages, and in the undiluted lysate there are 470 x 5 per milliliter (since 0.2 mL was plated), or 2350/mL. The r^+ phages were generated by recombination between the two *rII* mutations. The other product of the recombination event is the double mutant, and it does not grow on *K12*(λ). Therefore the true number of recombinants in the population is actually equivalent to twice the number of r^+ phages, since for every wild-type phage produced, there ought to be a doubly mutant recombinant produced. Thus there are 4700 recombinants/mL. The total number of phages in the lysates is 627 x (dilution factor) x (1 mL divided by the sample size plated) per milliliter, or 672 x 1000 x 10 = 6,720,000/mL. The map distance between the mutations is (4700/6,720,000) x 100 = 0.07%.

7.17   Construct a map from the following two factor phage cross data (show map distance):

Class	% Recombination
r_1 x r_2	0.10
r_1 x r_3	0.05
r_1 x r_4	0.19
r_2 x r_3	0.15
r_2 x r_4	0.10
r_3 x r_4	0.23

Answer:

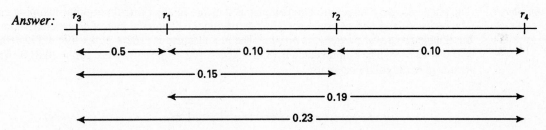

7.18   The following two-factor crosses were made to analyze the genetic linkage between four genes in phage λ: *c,mi,s,* and *co*.

Parents	Progeny
c + x + *mi*	1213 *c* +, 1205 + *mi*, 84 + +, 75 *c mi*
c + x + *s*	566 *c* +, 808 + *s*, 19 + +, 20 *c s*
co + x + *mi*	5162 *co* +, 6510 + *mi*, 311 + +, 341 *co mi*
mi + x + *s*	502 *mi* +, 647 + *s*, 65 + +, 56 *mi s*

Construct a genetic map of the four genes.

Answer: There are two answers that are compatible with the data:

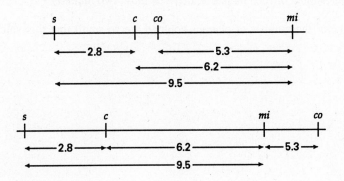

7.19   Three gene loci in T4 that affect plaque morphology in easily distinguishable ways are *r* (rapid lysis), *m* (minute), and *tu* (turbid). A culture of *E. coli* is mixedly infected with two types of phage *r m tu* and *r⁺ m⁺ tu⁺*. Progeny phage are collected and the following genotype classes are found:

Class			Number
r^+	m^+	tu^+	3,729
r^+	m^+	tu	965
r^+	m	tu^+	520
r	m^+	tu^+	172
r^+	m	tu	162
r	m^+	tu	474
r	m	tu^+	853
r	m	tu	3,467
			10,342

Construct a map of the three genes. What is the coefficient of coincidence, and what does the value suggest?

Answer: The phage progeny values are handled just as they are in any other three-point-mapping cross. The order of the genes, as established by identifying the reciprocal pair of classes that represent the double-crossover progeny, is *m-r-tu*. The *m-r* distance is (162 + 520 + 474 + 172)/10,342 x 100 = 12.84%. The *r-tu* distance is (853 + 162 + 172 + 965)/10,342 x 100 = 20.81%. These results give the following map:

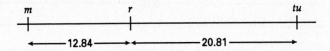

The observed double-crossover frequency is (162 + 172)/10,342 x 100 = 3.23%. The expected double-crossover frequency derived from the map distances just calculated is (0.1284 x 0.2081) x 100 = 2.67%. The coefficient of coincidence is 3.23/2.67 = 1.21. This value indicates the absence of interference in this cross. In fact, since the value is greater than one, we may hypothesize in this case that the presence of one crossover in a region actually enhances the occurrence of a second crossover nearby. This phenomenon is called negative interference.

7.20   The *rII* mutants of bacteria T4 grow in *E. coli B* but not in *E. coli K12*(λ) . The *E. coli* strain *B* is doubly infected with two *rII* mutants. A 6 x 10⁷ dilution of the lysate is plated on *E. coli B*. A 2 x 10⁵ dilution is plated on *E. coli K12*(λ). Twelve plaques appeared on strain *K12*(λ), and 16 on strain *B*. Calculate the amount of recombination between these two mutants.

Answer: 0.5% recombination (the number of plaques counted are so small that this value is a rough approximation)

7.21 Wild-type (r^+) strains of T4 produce turbid plaques, whereas *rII* mutant strains produce larger, clearer plaques. Five *rII* mutations (*a-e*) in the *A* cistron of the *rII* region of T4 give the following percentages of wild-type recombinants in two-point crosses:

Class	% of Wild-type Recombinants
a x b	0.2 percent
a x c	0.9 percent
a x d	0.4 percent
b x c	0.7 percent
e x a	0.3 percent
e x d	0.7 percent
e x c	1.2 percent
e x b	0.5 percent
b x d	0.2 percent
d x c	0.5 percent

What is the order of the mutational sites and what are the map distances between the sites?

Answer:

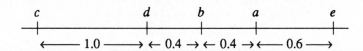

Note that the reciprocal recombinant class is the double *rII* mutant, which is indistinguishable from the single *rII* mutants in phenotype. Therefore, the frequencies of wild-type recombinants between sites are doubled to give map disatances.

7.22 Choose the correct answer in each case:
   a. If one wants to know if two different *rII* point mutants lie at exactly the same site (nucleotide pair), one should:
   i. Coinfect *E. coli K12(λ)* with both mutants. If phage are produced, they lie at the same site.
   ii. Coinfect *E. coli K12(λ)* with both mutants. If phage are not produced, they lie at the same site.
   iii. Coinfect *E. coli B* with both mutants and plate the progeny phage on both *E. coli B* and *E. coli K12(λ)*. If plaques appear on *B* but not *K12(λ)*, they lie at the same site.
   iv. Coinfect *E. coli K12(λ)* with both mutants and plate the progeny phage on both *E. coli B* and *E. coli K12(λ)*. If plaques appear on *K12(λ)* but not *B* they lie at the same site.
   b. If one wants to know if two different *rII* point mutants lie in the same cistron, one should:
   i. Coinfect *E. coli K12(λ)* with both mutants. If phage are produced, they lie at the same cistron.
   ii. Coinfect *E. coli K12(λ)* with both mutants. If phage are not produced, they lie in the same cistron.
   iii. Coinfect *E. coli B* with both mutants and plate the progeny phage on both *E. coli B* and *E.coli K12(λ)*. If plaques appear on *B* but not *K12(λ)*, they lie in the same cistron.
   iv. Coinfect *E. coli K12(λ)* with both mutants and plate the progeny phage on both *E. coli B* and *E. coli K12(λ)*. If plaques appear on *K12(λ)* but not *B* they lie in the same cistron.

Answer: a. iii
          b. ii

7.23 Given the following map with point mutants, and given the data in the table below, draw a topological representation of deletion mutants *r21, r22, r23, r24* and *r25*. (Be sure to clearly the end points of the deletions.)

($+ = r^+$ recombinants are obtained. $0 = r^+$ recombinants are not obtained.)

Map:

r12   r16   r11   r15   r13   r14   r17

Deletion Mutants	Point mutants						
	r11	*r12*	*r13*	*r14*	*r15*	*r16*	*r17*
r21	0	+	0	+	0	+	+
r22	+	+	0	0	+	+	0
r23	0	0	0	+	0	0	+
r24	+	+	0	0	+	+	+
r25	+	+	0	0	0	+	+

Answer:

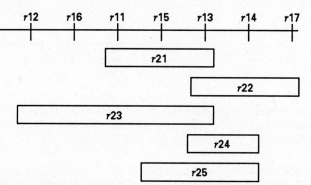

7.24. A set of seven different *rII* deletion mutants of bacteriophage T4, *1* through *7*, were mapped, with the following result:

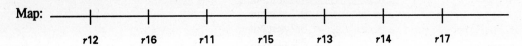

Five *rII* point mutants were crossed with each of the deletions, with the following results, where + = r^+ recombinants, were obtained, 0 = no r^+ recombinants were obtained:

Point Mutants	Deletion Mutants						
	1	2	3	4	5	6	7
a	0	+	+	+	0	0	0
b	0	0	+	+	+	+	0
c	0	+	+	0	0	0	+
d	0	+	0	0	+	0	+
e	0	+	+	+	+	0	0

Map the locations of the point mutants.

Answer:

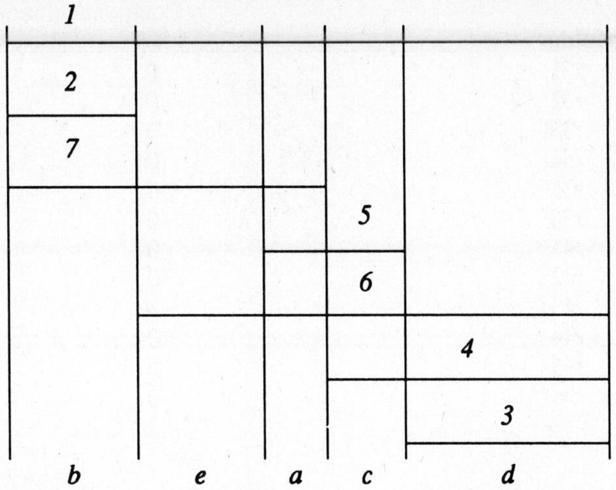

7.25 Given the following deletion map with deletions r31, r32, r33, r34, r35 and r36, place the point mutants r41, r42 etc., on the map. Be sure you show where they lie with respect to end points of the deletions.

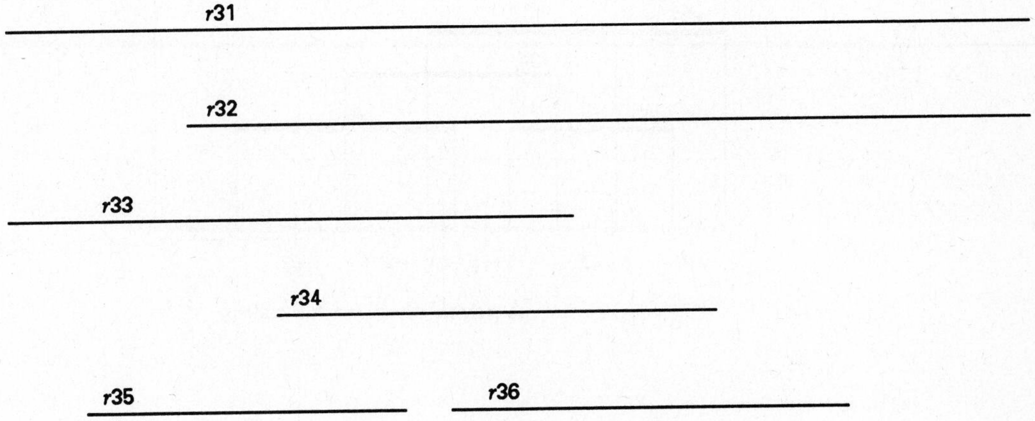

r31

r32

r33

r34

r35   r36

Point Mutants	Deletion mutants (+ = recombinants produced; 0= No r^+ recombinants produced)					
	r31	r32	r33	r34	r35	r36
r41	0	0	0	0	+	0
r42	0	0	0	+	0	+
r43	0	0	+	+	+	0
r44	0	0	0	0	+	+
r45	0	+	0	+	+	+
r46	0	0	+	0	+	0

Show the dividing line between the *A* cistron and the *B* cistron on your map above from the following data: [+ = growth on strain *K12*(λ), 0 = no growth on strain *K12*(λ)]:

Mutant	Complementation with	
	rIIA	*rIIB*
r31	0	0
r32	0	0
r33	0	+
r34	0	0
r35	0	+
r36	0	0
r41	0	+
r42	0	+
r43	+	0
r44	0	+
r45	0	+
r46	0	+

Answer:

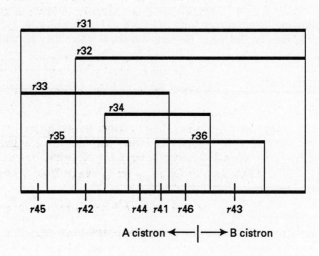

7.26   Mutants in the *ade2* gene of yeast require adenine and are pink because of the intracellular accumulation of a red pigment. Diploid strains were made by mating haploid mutant strains. The offspring exhibited the following phenotypes:

Cross	Diploid Phenotypes
1 x 2	pink, adenine-requiring
1 x 3	white, prototrophic
1 x 4	white, prototrophic
3 x 4	pink, adenine-requiring

How many genes are defined by the four different mutants? Explain?

Answer: If two mutations are in the same unit of function (i.e. the same gene), then they will not complement when together in a diploid cell and the mutant phenotype will be exhibited. If two mutations are in different genes, they will complement a diploid cell and the wild-type phenotype will result. The data presented for diploids made between pairs of four *ade2* mutants indicate that 1 and 2, and 3 and 4 do not complement, while 1 and 3, and 1 and 4 do complement. Thus, there are two genes: mutations 1 and 2 are in one gene, and mutations 3 and 4 are in the other gene.

7.27 In *Drosophila* mutants *A, B, C, D, E, F,* and *G* all have the same phenotype: the absence of red pigment in the eyes. In pairwise combinations in complementation tests the following results were produced, where + = complementation and − = no complementation.

	A	B	C	D	E	F	G
G	+	−	+	+	+	+	−
F	−	+	+	−	+	−	
E	+	+	−	+	−		
D	−	+	+	−			
C	+	+	−				
B	+	−					
A	−						

a. How many genes are present?
b. Which mutants have defects in the same gene?

Answer: a. 3 genes
b. *A, D,* are *F* are in one; *B* and *G* are in the second; and *C* and *E* are in the third.

7.28 a. A homozygous white eyed Martian fly (w_1/w_1) is crossed to a homozygous white eyed fly from a different stock (w_2/w_2). It is well known that wild-type Martian flies have red eyes. This cross produces all white eyed progeny. State whether the following is true or false. Explain your answer.
i) w_1 and w_2 are allelic genes.
ii) w_1 and w_2 are non-allelic.
iii) w_1 and w_2 affect the same function.
iv) The cross was a complementation test.
v) The cross was a *cis-trans* test.
vi) w_1 and w_2, are allelic by terms of the functional test.
The F_1 white eyed flies were allowed to interbreed, and when you classified the F_1 you found 20,000 white-eyed flies and ten red-eyed progeny. Concerned about contamination, you repeat the experiment and get exactly the same results. How can you best account for the presence of the red-eyed progeny? As part of your explanation give the genotypes of the F_1 and F_2 generation flies.

Answer: i) The two mutations did not complement.
ii) False. Non-allelic genes would have complemented and red-eyed flies would have been produced.
iii) True. Both mutations are in the same gene.
iv) True. As explained above.
v) True. This was the *trans* configuration.
vi) True. If two genes affect the same function, they are allelic.

b. The P cross was $\dfrac{w_1}{w_1}$ x $\dfrac{w_2}{w_2}$ → F_1 $\dfrac{w_1}{w_1}$ (white).

However, apparently crossover occurred in the F_1. So we can re-write the genotype.

$$\frac{w_1\,+}{w_1\,+} \text{ x } \frac{+\,w_2}{+\,w_2}$$

So the F_1 heterozygote really is $\dfrac{w_1\,+}{+\,w_2}$ (white).

Since w_1 and w_2 involve different nucleotides, crossing-over can occur and be detected. But since both of these mutations are within the same gene, they are close together and so the frequency of crossover is very rare. The events can now be represented as follows:

In the F$_1$ cross $\dfrac{w_1 \pm}{+ \; w_2}$ x $\dfrac{w_1 \pm}{+ \; w_2}$

if no crossover occurs, all progeny would be white. But if a crossover rarely occurs

$$\dfrac{w1 \quad +}{ \times }$$
$$+ \quad w2$$
$$\longrightarrow \quad \underline{+ \quad +} \quad \text{and} \quad \underline{w1 \;\; w2}$$

$\underline{\pm \pm}$ and $\underline{w_1 \; w_2}$ gametes are produced. The $\underline{w_1 \; w_2}$ gamete would not be detected but a gamete $\underline{\pm \pm}$ would determine red-eyed progeny. So the answer is that intragenic recombination has occurred in the trans heterozygote. (This is analogous to Benzer's *cis-trans* test in *rII*-region of T4.)

CHAPTER 8

THE BEGINNINGS OF MOLECULAR GENETICS: GENE FUNCTION

I. CHAPTER OUTLINE

Gene Control of Enzyme Structure
    Garrod's Hypothesis of Inborn Errors of Metabolism
    Genetic Control of *Drosophila* Eye Pigments
    One Gene-One Enzyme Hypothesis
Genetically Based Enzyme Deficiencies in Humans
    Phenylketonuria
    Albinism
    Lesch-Nyhan Syndrome
    Tay-Sachs Disease
    Genetic Counseling
Gene Control of Protein Structure
    Protein Structure
    Sickle–Cell Anemia
    Other Hemoglobin Mutants
    Biochemical Genetics of the Human ABO Blood Groups

II. IMPORTANT TERMS AND CONCEPTS

alkaptonuria	polypeptide
phenylketonuria (PKU)	amino acid
autonomous and	amino group
nonautonomous development	carboxyl group
Lesch–Nyhan syndrome	α-carbon
Tay-Sachs disease	radical (R) group
lysosomal-storage disease	peptide bond
amniocentesis	protein structure (1°-4°)
chorionic villus sampling	α–helix
biochemical pathway	sickle-cell anemia

III. THINKING ANALYTICALLY

Solving this set of problems, some of which are concerned with identifying the steps in a biosynthetic pathway with particular genes, is a bit like detective work in which the object is to disclose how the "crime" was committed, the location of the scene of the crime, and what accessories before or after the fact, if any, were involved. The "criminals" in these situations are mutant alleles, and the actual perpetrators of the crimes, if any, are enzymes.

    The solutions depend upon, first, a well planned investigation of the facts; next, preparation of a list of suspects; then a description of the sequence of events leading up to the crime; and, last, identification of the miscreants.

QUESTION: The police rounded up the Hunker family, suspects in a series of petty thefts. The investigation came up with the following: Oscar Hunker stole half as much as his mother, Mary. Mary stole five times as much as her daughter, Judy, and $75 more than her son, Nick. Judy pinched ten times as much as her little sister, Agnes, who tried very hard. Agnes managed to get $5 for the candy bars she stole. Jack, the father, was charged with grand larceny, because he stole twice as much as the rest of the family combined.

Rank the members of the family according to amount stolen, showing the amount in each case.

Solution: In order to solve this problem, it is necessary at some point to establish a monetary base. The only clue is the $5 that little Agnes got. Proceed from there.

If Agnes got $5, then Judy stole $50, and Mary, who stole 5 times as much as Judy, stole $250. Nick stole $175 ($75 less than Mary), and Oscar, who stole half as much as his mother, gets credit for having stolen $125. The total so far is $605. Twice this, $1210, is the amount Jack stole.

Answer: Jack $1210, Mary $250, Nick $175, Oscar $125, Judy $50, and Agnes $5. Total = $1815.00.

IV. QUESTIONS FOR PRACTICE

A. Multiple Choice Questions

1.   Archibald Garrod is noted for his studies of the inheritance of
      a.   sickle-cell anemia.
      b.   *Drosophila* eye color.
      c.   alkaptonuria.
      d.   phenylketonuria.

2.   Beadle & Tatum established the principle known as
      a.   the theory of genetic relativity.
      b.   the theory of autonomous development.
      c.   the theory of the α–helix.
      d.   the one gene-one enzyme hypothesis.

3.   A major symptom of PKU is the accumulation in the blood and tissues of the compound
      a.   homogentisic acid.
      b.   phenylketones.
      c.   paradichlorbenzene.
      d.   all of the foregoing

4.   The recessive allele responsible for albinism in humans interferes with the biosynthetic pathway leading from
      a.   tyrosine to melanin.
      b.   inosine to adenosine.
      c.   hypoxanthine.
      d.   tyrosine to homogentisic acid.

5.   The Lesch-Nyhan syndrome is attributable to the accumulation of excess
      a.   HGPRT.
      b.   purines and uric acid.
      c.   phenylalanine.
      d.   lysosomal enzymes.

6.   Tay-Sachs disease is characterized by
      a.   rapid degeneration of brain function.
      b.   generalized paralysis.
      c.   blindness.
      d.   all of the foregoing

92

7.   Polypeptides are polymers of
    a.   peptidases.
    b.   amino acids.
    c.   proteins.
    d.   none of the foregoing

8.   Amino acids are characterized by the possession of
    a.   a carboxyl group.
    b.   a bicarbonate group.
    c.   tertiary structure.
    d.   a ketone group.

9.   The covalent bonds that join amino acids together to form larger molecules are
    a.   hydrogen bonds.
    b.   peptide bonds.
    c.   ionic bonds.
    d.   carbonyl bonds.

10.   Quaternary protein structure refers to the complex of two or more polypeptide chains that comprise
    a.   the R-groups of the amino acids.
    b.   the peptide bonds.
    c.   the α-helix of certain compounds.
    d.   certain protein molecules.

Answers: 1c, 2d, 3b, 4a, 5b, 6d, 7b, 8a, 9b, 10d

B.   Thought Questions

1.   If you were a genetic counselor, how would you counsel this couple, who are contemplating marriage? The prospective groom is apparently normal, except for a nervous twitch of his nose, but his maternal uncle had PKU as a child. The prospective bride is also apparently normal, but her father has hemophilia and her brother plays in a rock band. (cf. Text pp. 248-250)

2.   Distinguish between the procedures of amniocentesis and chorionic villus sampling, and comment on their usefulness in genetic counseling. (cf. Text pp. 249-250)

3.   Discuss the circumstances encountered by Beadle and Ephrussi in their investigations of the inheritance of eye color in *Drosophila*. (cf. Text pp. 235-238)

4.   Discuss the implications of inherited disorders, such as sickle-cell anemia, with respect to evolution by natural selection. (cf . Text pp. 256-257)

5.   A group of mutants that require one of the following nutrient supplements for growth was tested, with the following results. (+ = growth, - = no growth)

	Nutrients			
Mutant	W	X	Y	Z
1	–	–	–	+
2	+	+	–	+
3	–	+	–	+

What is the order of nutrients in the biosynthetic pathway of these nutrients, and at what points in the pathway does each mutant effect a block?

Solution: All mutants will grow if Z is added to the minimum medium, so Z must be the end point. Mutant 3 will grow in the presence of either Z or X, but not with W as the only supplement. Mutant 2 requires only W, but once it has it, it does not need either X or Z. Y is the precursor, present in the minimum medium.

Looking at it in another way:

a.   Mutation 2 blocks the conversion of precursor Y to intermediate W. So the normal allele of mutant 2 must code the synthesis of enzyme 2, which is needed to catalyse this conversion. If W is provided, the biosynthetic pathway can resume and complete the synthesis of X and Z, and the organism will grow.

b. Mutation 3 blocks the conversion of W to X, which is essential to the formation of Z. If X is added to the medium for this mutant strain, the pathway can resume and be completed with substance Z. So the wild type allele of 3 codes the enzyme that normally catalyses this reaction.

c. Mutation 1 prevents conversion of X to Z, the lack of which prevents growth of this mutant strain. Substance Z must be added to the culture medium if growth is to take place. But as long as Z is provided, neither X nor W need be added to the culture medium, although their presence will not inhibit growth. The wild type allele of mutant 1 normally assures the X to Z conversion.

The answer is:

$$\overset{2}{Y} \rightarrow \overset{3}{W} \rightarrow \overset{1}{X} \rightarrow Z$$

6. Similar to the hypothetical example in Problem 5 are the data on the biosynthesis of the amino acid methionine. Three separate mutant strains of *Neurospora* that were methionine auxotrophs were tested. One strain (a) required the addition of methionine in order to grow. A second (b) required either methionine or homocysteine, and the third (c) required either methionine or homocysteine or cystathione. Using + for growth and − for no growth, complete the diagram below, correlating the three mutations with the nutritional supplement(s) that they require or will accept for growth. Also show the biosynthetic pathway and the location of the mutation blocks.

Mutant Strains	Precursor	Cystathione	Methionine	Homocysteine
a				
b				
c				

Solution: Mutant strain (a) has methionine as an absolute requirement, and will not grow with either homocysteine or cystathione as the only supplement. Strain (b) will grow with either methionine or homocysteine, and strain (c) will grow with either methionine or homocysteine or cystathione as a supplement. The diagrams therefore would be:

Mutant Strains	Precursor	Cystathione	Methionine	Homocysteine
a	−	−	+	−
b	−	−	+	+
c	−	+	+	+

$$\text{Precursor} \rightarrow \underset{c}{\text{Cys}} \rightarrow \underset{b}{\text{Hom}} \rightarrow \underset{a}{\text{Met}}$$

7. Describe and distinguish between the primary, secondary, tertiary, and quarternary structure of proteins, giving examples of each. (cf Text pp. 251-253, Fig. 8.21)

8. Describe the functions of genetic counseling services - what they look for, how they proceed, what they recommend, and what, if any, advice they give. (cf. Text pp. 248-250)

V. ANSWERS AND SOLUTIONS TO TEXT QUESTIONS

8.1   Phenylketonuria (PKU) is a heritable metabolic disease of humans; its symptoms include mental deficiency. The gross phenotypic effect is due to:
   a. Accumulation of phenylketones in the blood
   b. Accumulation of maple sugar in the blood
   c. Deficiency of phenylketones in the blood
   d. Deficiency of phenylketones in the diet

Answer: a. Accumulation of phenylketones in the blood

8.2   If a person were homozygous for both PKU (phenylketonuria) and AKU (alkaptonuria), what would you expect his or her phenotype to be? Refer to the pathway below.

Phenylalanine
↓ (blocked in PKU)
tyrosine → DOPA → melanin
↓
p-Hydroxyphenylpyruvic acid
↓
Homogentisic acid
↓ (blocked in AKU)
Maleylacetoacetic acid

Answer: The double homozygote should have PKU but not AKU. The PKU block should prevent most homogentistic acid from being formed, so that it could not accumulate to high levels.

8.3   Refer to the pathway shown in 8.2. What effect, if any, would you expect PKU (phenylketonuria) and AKU (alkaptonuria) to have on pigment formation?

Answer: PKU should lead to a decrease in tyrosine levels (although some tyrosine is obtained from food), and thus to a decrease in melanin formation. Indeed PKU patients are reported sometimes to be lighter in pigmentation than their normal relatives. AKU, if it has an effect, might cause elevation of tyrosine levels due to downstream blockage of the pathway. This could yield increased melanin.

8.4 a^+, b^+, c^+, and d^+ are independently assorting Mendelian genes controlling the production of a black pigment. The alternate alleles that give abnormal functioning of these genes are a, b, c, and d. A black a^+/a^+ b^+/b^+ c^+/c^+ d^+/d^+ is crossed with a colorless a/a b/b c/c d/d to give a black F_1. The F_1 is then selfed. Assume that a^+, b^+, c^+, and d^+ act in a pathway as follows:

$$\begin{array}{cccc} a^+ & b^+ & c^+ & d^+ \end{array}$$
colorless → colorless → colorless → brown → black

    a.   What proportion of the F_2 are colorless?
    b.   What proportion of the F_2 are brown?

Answer: a.   The simplest approach is to calculate the proportion of the F_2 that are colored and to subtract that answer from one. In this case the noncolorless are the brown or black progeny. The proportion of progeny that make at least the brown pigment is given by the probability of having the following genotype: $a^+/- b^+/- c^+/- (d^+/d^+, d^+/d \text{ or } d/d)$. The answer is 3/4 x 3/4 x 3/4 x 1 = 27/64. Therefore the proportion of colorless is 1 - 27/64 = 37/64.

        b.   The brown progeny have the following genotype: $a^+/- b^+/- c^+/- d/d$. The probability of getting individuals with this genotype is 3/4 x 3/4 x 3/4 x 1/4 = 27/256.

8.5   Using the genetic information given in Problem 8.4, now assume that a^+, b^+, and c^+ act in a pathway as follows:

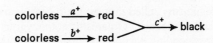

Black can be produced only if both red pigments are present, i.e. C converts the two red pigments together into a black pigment.

    a. What proportion of the F_2 are colorless?

    b. What proportion of the F_2 are red?

    c. What proportion of the F_2 are black?

Answer: a. To be colorless, an individual must be a/a b/b ($c^+/-$ or c/c). The probability of this genotype is $1/4 \times 1/4 \times 1 = 1/16$.

    b. To be red, an individual must be either $a^+/-$ or $b^+/-$ or both, and it must be c/c. With regard to the a and b loci, it is easier to calculate the probability of colorless, that is, a/a b/b, and subtract that value from one to get the probability of obtaining red color with just those two loci. The probability of colorless is $1/4 \times 1/4 = 1/16$; therefore, the probability of red is $1 - 1/16 = 15/16$. Now considering the $1/4$ chance of being c/c and hence not converting red to black, the overall proportion of the F_2 that are red is $15/16 \times 1/4 = 15/64$.

    c. To be black, an individual must be $a^+/- b^+/- c^+/-$, and the proportion of individuals that have this genotypic constitution is $3/4 \times 3/4 \times 3/4 = 27/64$.

8.6   a. Three genes on different chromosomes are responsible for three enzymes that catalyze the same reaction in corn:

$$a^+,b^+,c^+$$
$$\text{colorless compound} \quad \rightarrow \quad \text{red compound}$$

The normal functioning of any one of these genes is sufficient to convert the colorless compound to the red compound. The abnormal functioning of these genes is designated by a, b, and c, respectively. A red a^+/a^+ b^+/b^+ c^+/c^+ is crossed by a colorless a/a b/b c/c to give a red F_1, a^+/a b^+/b c^+/c. The F_1 is selfed. What proportion of the F_2 are colorless?

b. It turns out that another step is involved in the pathway. It is controlled by gene d^+, which assorts independently of a^+, b^+, and c^+:

$$d^+ \qquad\qquad\qquad a^+, b^+, c^+$$
$$\text{colorless compound 1} \rightarrow \text{colorless compound 2} \rightarrow \text{red compound}$$

The inability to convert colorless 1 to colorless 2 is designated d. A red a^+/a^+ b^+/b^+ c^+/c^+ d^+/d^+ is crossed by a colorless a/a b/b c/c d/d. The F_1 are all red. The red F_1's are now selfed. What proportion of the F_2 are colorless?

Solution: a. Since any one of the normal alleles (a^+, b^+, c^+) is capable of catalyzing the reaction leading to color, in order for color to fail to develop all three must be deficient – that is, the genotype must be a/a b/b c/c. The probability of being homozygous for each of these recessive alleles is $1/4$, and for all three is $1/4^3 = 1/64$.

b. If the probability of being colorless, therefore, is $1/64$, then the probability of being colored is $1 - 1/64 = 63/64$. The probability of having the d^+ allele (d^+/d^+ or d^+/d) = $3/4$, so the combined probability = $3/4 \times 63/64 = 189/256$. Therefore the remainder ($256/256 - 189/256 = 67/256$) will be colorless.

Answer: a. $1/64$

    b. $67/256$

8.7   In hypothetical organisms called mongs, the recessive bw causes a brown eye, and the (unlinked) recessive st causes a scarlet eye. Organisms homozygous for both recessives have white eyes. The genotypes and corresponding phenotypes, then, are as follows:

$$bw^+/- \quad st^+/- \quad \text{red eye}$$
$$bw/bw \quad st^+/- \quad \text{brown}$$
$$bw^+/- \quad st/st \quad \text{scarlet}$$
$$bw/bw \quad st/st \quad \text{white}$$

Outline a hypothetical biochemical pathway that would give this type of gene interaction. Demonstrate why each genotype shows its specific phenotype.

Answer: The simplest scheme is one in which there are two distinct biochemical pathways, one producing a brown pigment and the other producing a scarlet pigment. See the diagram. The two pigments then combine, or are combined, to produce a red pigment that gives the eye its color. Individuals homozygous for *bw* have a defective enzyme in the scarlet pigment pathway so that only brown pigment is produced, that is, they are scarletless. Similarly, individuals homozygous for *st* have a defective enzyme in the brown pathway so that only scarlet pigment is produced; that is, they are brownless. Double mutants are scarletless and brownless, giving a white eye.

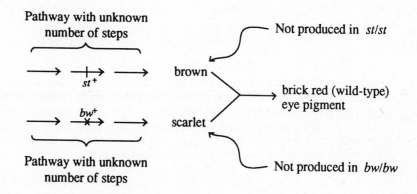

8.8 The Black Riders of Mordor in *Lord of the Rings* ride steeds with eyes of fire. As a geneticist you are very interested in the inheritance of the fire red eye color. You discover that the eyes contain two types of pigments, brown and red, that are usually bound to core granules in the eye. In wild-type steeds precursors are converted by these granules to the above pigments, but in steeds homozygous for the recessive X-linked gene *w* (white eye), the granules remain unconverted and a white eye results. The metabolic pathways for the synthesis of the two pigments are shown in the following figure. Each stop of the pathway is controlled by a gene: A mutation *v* gives vermilion eyes; *cn* gives cinnabar eyes; *st* gives scarlet eyes; *bw* gives brown eyes; and *se* gives black eyes. All the mutations are recessive to their wild-type alleles and all are unlinked. For the following genotypes, show the phenotypes and proportions of steeds that would be obtained in the F_1 of the given matings.

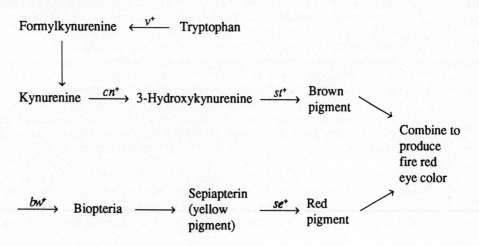

97

a. w/w bw^+/bw^+ st/st x w^+/Y bw/bw st/st

b. w^+/w^+ se/se bw/bw x w/Y se^+/se^+ bw^+/bw^+

c. w^+/w^+ v^+/v^+ bw/bw x w/Y v/v bw/bw

d. w^+/w^+ bw^+/bw st^+/st x w/Y bw/bw st/st

Answer:
  a. Half have white eyes (the sons) and half have fire red eyes (the daughters).

  b. All have fire red eyes.

  c. All have brown eyes.

  d. All are w^+; a fourth are $bw^+/-$ $st^+/-$, red; a fourth are $bw^+/-$ st/st, scarlet; a fourth are bw/bw $st^+/-$, brown; a fourth are bw/bw st/st, the color of 3-hydroxykyrurenine plus the color of the precursor to biopterin, or colorless.

8.9   Upon infection of *E. coli* with bacteriophage T4, a series of biochemical pathways result in the formation of mature progeny phages. The phages are released following lysis of the bacterial host cells. Let us suppose that the following pathway exists:

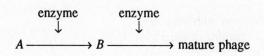

Let us also suppose that we have two temperature-sensitive mutants that involve the two enzymes catalyzing these sequential steps. One of the mutations is cold-sensitive (*cs*) in that no mature phages are produced at 17°C. The other is heat-sensitive (*hs*) in that no mature phages are produced at 42°C. Normal progeny phages are produced when phages carrying either of the mutations infect bacteria at 30°C. However, let us assume that we do not know the sequence of the two mutations. Two models we therefore two models are therefore apparent:

$$hs \quad cs$$
(1) A $\rightarrow$ B $\rightarrow$ phage

$$cs \quad hs$$
(2) A $\rightarrow$ B $\rightarrow$ phage

Outline how you would experimentally determine which model is the correct model without artificially breaking phage-infected bacteria.

Answer: Wild-type T4 will produce progeny phages at all three temperatures. Let us suppose that model (1) is correct. If cells infected with the double mutant are first incubated at 17°C and then shifted to 42°C, then progeny phages will be produced and the cells will lyse. The explanation is as follows: The first step, *A* to *B*, is controlled by a gene whose product is heat-sensitive. At 17°C the enzyme works and *A* is converted to *B*, but *B* cannot then be converted to mature phages since that step is cold-sensitive. When the temperature is raised to 42°C, the *A*-to-*B* step is now blocked, but the accumulated *B* can be converted to mature phages since the enzyme involved with that step is cold-sensitive, and so the enzyme is functional at the high temperature. If model 2 is the correct pathway, then progeny phages should be produced in a 42°-to-17°C temperature shift, but not vice versa. In general, two gene product functions can be ordered by this method whenever one temperature shift allows phage production and the reciprocal shift does not, according to the following rules: (1) If a low-to-high temperature results in phages but a high-to-low temperature does not, then the *hs* step precedes the *cs* step (model 1); (2) If a high-to-low temperature results in phages but a low-to-high temperature does not, then the *cs* step precedes the *hs* step (model 2).

8.10   Four mutant strains of *E. coli* (*a, b, c,* and *d*) all require substance X in order to grow. Four plates were prepared, as shown in the following figure. In each case the medium was minimal, with just a trace amount of substance X, to allow a small amount of growth of mutant cells. On plate *a,* cells of mutant strain *a* were spread over the agar, and grew to form a thin lawn. On plate *b* the lawn is composed of mutant *b* cells, and so on. On each plate, cells of the four mutant types were inoculated over the lawn, as indicated by the circles. Dark circles indicates luxuriant growth occurred. That is, this experiment tests whether the bacterial strain spread on the plate can "feed" the four strains inoculated on the plate, allowing them to grow. What do these results show about the relationship of the four mutants to the metabolic pathway leading to substance X?

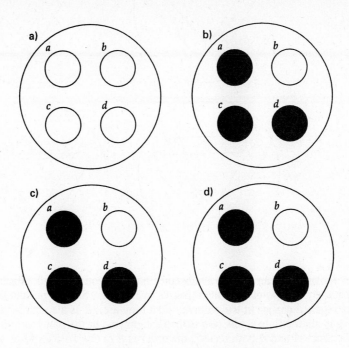

Answer: Mutant *a* blocks the earliest step in the pathway as it cannot feed any of the others. Mutant *c* is next, as it can supply the substance *a* needs, but cannot food *b* or *d*. Mutant *d* is next, and mutant *b* is the last in the pathway, as it can feed all the others.

8.11   The following table indicates what enzyme is deficient in six different complementing mutants of *E. coli*, none of which can grown on minimal medium. All of them will grow if tryptophan (Trp) is added to the medium.

Mutant	Enzyme Missing
trpE	anthranilate synthetase
trpA	tryptophan synthetase
trpF	IGP synthetase
trpB	tryptophan synthetase
trpD	PPA transferase
trpC	PRA isomerase

Each of the plates in the following figure shows the results of streaking three of the mutants on minimal medium with just a trace of added tryptophan. Heavy shading indicates regions of heavy growth, indicating where a strain can be fed by the strain streaked next to it on the plate. In what order do the enzymes listed above act in the tryptophan synthetic pathway?

99

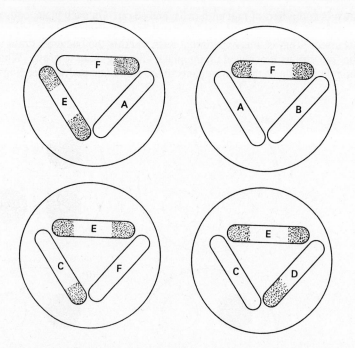

Answer: A mutant cell can take up a substance and grow on it only if substance occur in the metabolic pathway at a point after the mutant's own block. Since *trpE* can be fed by *C, D, F,* and *A, E* is blocked earlier in the pathway than they are. *E* is also earlier than B since *B* can feed *F*. Thus anthranilate synthetase is the first enzyme in the pathway. *F* can feed *C* and *C* can feed *D*. Since *A* and *B* both feed *F*, the order is *E D C F* [*A B*]. After anthranilate synthetase, the enzymes are, in order, PPA transferase, PRA isomerase, IGP synthetase and tryptophan synthetase.

8.12   Referring to the list of mutants and enzymes given in Problem 8.11, explain how it can be that two different complementing mutants (*trpA* and *trpB*) can affect the activity of the same enzyme. How will two such mutants be related in terms of their position within the metabolic pathway?

Answer: Since *trpA* and *trpB* complement, they affect different polypeptides. Since they both affect typtophan synthetase, this enzyme must contain two (at least) different polypeptide chains. The position in the metabolic pathway is purely a function of which enzyme is affected (which catalytic step). Thus *trpA* and *trpB* affected the same step of the pathway.

8.13   Two mutant strains of *Neurospora* lack the ability to make compound Z. When crossed, the strains usually yield asci of two types: (1) those with spores that are all mutant and (2) those with four wild-type and four mutant spores. The two types occur in a 1:1 ratio.
    a.   Let *c* represent one mutant, and let *d* represent the other. What are the genotypes of the two mutant strains?
    b.   Are *c* and *d* linked?
    c.   Wild-type strains can make compound Z from the constituents of the minimal medium. Mutant *c* can make Z if supplied with X but not if supplied with Y, while mutant *d* can make Z from either X or Y. Construct the simplest linear pathway of the synthesis of Z from the precursors X and Y, and show where the pathway is blocked by mutations *c* and *d*.

Answer: a. $c\ d^+$ and $c^+\ d$
    b.   The genes are not linked since parental ditype (PD) and nonparental ditype (NPD) tetrads occur in equal frequencies.
    c.   The pathway is Y to X to Z, with *d* blocking the synthesis of Y and *c* blocking the synthesis of X from Y.

8.14 The following growth responses (where + = growth and 0 = no growth) of mutants *1-4* were seen on the related biosynthetic intermediates A, B, C, D, and E. Assume all intermediates are able to enter the cell, that each mutant carries only one mutation, and that all mutants affect steps after B in the pathway.

	Growth on				
Mutant	A	B	C	D	E
1	+	0	0	0	0
2	0	0	0	+	0
3	0	0	+	0	0
4	0	0	0	+	+

Which of the schemes in the following figure fits best with the data with regard to the biosynthetic pathway?

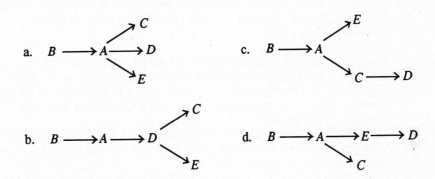

Answer: The scheme in part d.

8.15 Four strains of *Neurospora*, all of which require arginine but have unknown genetic constitution, have the following characteristics. The nutrition and accumulation characteristics are as follows:

	Growth On				
Strain	Minimal Medium	Ornithine	Citrulline	Arginine	Accumulates
1	–	–	+	+	Ornithine
2	–	–	–	+	Citrulline
3	–	–	–	+	Citrulline
4	–	–	–	+	Ornithine

The pairwise complementation tests of the four strains gave the following results (+ = growth on minimal medium and 0 = no growth on minimal medium):

	4	3	2	1
1	0	+	+	0
2	0	0	0	
3	0	0		
4	0			

Crosses among mutants yielded prototrophs in the following percentages:

1 x *2*: 25 percent
1 x *3*: 25 percent
1 x *4*: none detected among 1 million ascospores
2 x *3*: 0.002 percent
2 x *4*: 0.001 percent
3 x *4*: none detected among 1 million ascospores

Analyze the data and answer the following questions.
a. How many distinct mutational sites are represented among these four strains?
b. In this collection of strains, how many types of polypeptide chains (normally found in the wild type) are affected by mutations?
c. Give the genotypes of the four strains, using a consistent and informative set of symbols.
d. Give the map distances between all pairs of linked mutations.
e. Give the percentage of prototrophs that would be expected among ascospores of the following types: (1) strain *1* x wild type; (2) strain *2* x wild type; (3) strain *3* x wild type; (4) strain *4* x wild type.

Answer: Analysis of data: The nutrition and accumulation data suggest that strain *1* is blocked between orn and cit, because it accumulates the former and grows on the latter. Strains 2 and 3 are blocked, by the same reasoning, between cit and arg. Strain 4 is rather strange, since it has the growth properties of strains *2* and *3* and the accumulation phenotype of strain *1*. On these grounds, it is doubtful that it is a single mutant.

Complementation indicates that mutants 2 and 3 carry mutations in the same gene. That gene is not the same as in strain *1*. Strain 4 looks like a double mutant or a deletion mutant lacking two adjacent genes.

The question about strain 4 is resolved by the genetic tests. While mutations in strains *1* and *2* assort independently (25 percent prototrophs), the matings *1* x *4* and *3* x *4* yield no prototrophs. All data are consistent with the notion that strain *4* carries the mutation found in strain *1* *and* the distinct mutation found in strain *3*. (A mating of strain *4* x wild type should therefore give only 25 percent wild-type, instead of the 1:1 segregation expected of single mutants.)

Consistent with the allelism of mutations in strains 2 and 3 (note complementation) is their tight linkage in crosses. They are at different mutational sites, 0.004 map unit apart, but in the same gene. The remaining question about the difference in prototroph yield in the *2* x *3* cross and the *2* x *4* cross is deferred until answers to the specific questions are given.

a. 3
b. 2
c.

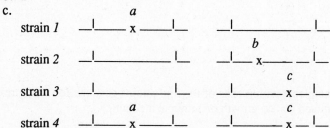

Two chromosomes are shown, with the ends of the relevant genes indicated. Mutations *b* and *c* are at distinct sites within one gene.
d. Mutations *b* and *c* are 0.004 map unit apart.
e. (i), (ii), (iii) Strains *1*, *2*, or *3*, when mated with wild type, will give 50 percent prototrophs and 50 percent auxotrophs. (iv) Strain *4*, as noted above, will give 25 percent prototrophs, since only one of the four equally frequent genotypes expected (i.e., $a^+ c^+$ from the cross $a^+ c^+$ x *a c*) will grow without arginine.

8.16 Proteins are (choose the correct answer):
    a. Branched chains of nucleotides
    b. Linear, folded chains of nucleotides
    c. Linear, folded chains, of amino acids
    d. Invariable enzymes

Answer: c

8.17. A breeder of Irish setters has a particularly valuable show dog that he knows is descended from the famous bitch Rheona Didona, who carried a recessive gene for atrophy of the retina. Before he puts the dog to stud, he must ensure that it is not a carrier for this allele. How should he proceed?

Answer: He should testcross the male to a retinal atrophic female. Retinal atrophic pups would indicate that the male is heterozygous.

8.18 Suppose you were on a jury to decide the following case. What would you conclude?
    The Jones family claims that Baby Jane, given to them at the hospital, does not belong to them but to the Smith family, and that the Smith's baby Joan really belongs to the Jones's family. It is alleged that the two babies were accidentally exchanged soon after birth. The Smiths deny that such an exchange has been made. Blood group determinations show the following results:

                 Jones mother, AB
                 Jones father, O
                 Smith mother, A
                 Smith father, O
                 Baby Jane, A
                 Baby Joan, O

Answer: Baby Joan, whose blood type is O, must belong to the Smith family. Mrs. Jones, being of blood type AB, could not have an O baby. Therefore Baby Jane must be hers.

8.19 Referring back to Figure 8.25, what would you expect the phenotype to be in individuals heterozygous for the following two hemoglobin mutations?
    a. Hb Norfolk and HbS
    b. HbC and HbS

Answer: a. The phenotype should be normal, as the two mutations will complement, and both normal α chains and normal β chains will be present.
               b. No normal β chains will be present, so there will be a mutant phenotype (anemia), though not as severe as in an HbS homozygote.

8.20 In evaluating my teacher, my sincere opinion is that:
    a. He (she) is a swell person whom I would be glad to have as a brother-in-law/sister-in-law
    b. He (she) is an excellent example of how tough it is when you don't have either genetics or environment going for you
    c. He (she) may have okay DNA to start with, but somehow all the important genes got turned off
    d. He (she) ought to be preserved in tissue culture for the benefit of other generations

Answer: Your choice!!

CHAPTER 9

THE STRUCTURE OF GENETIC MATERIAL

I. CHAPTER OUTLINE

The Nature of Genetic Material: DNA and RNA
    The Discovery of DNA as Genetic Material
    The Discovery of RNA as Genetic Material
The Chemical Composition of DNA and RNA
    The Physical Structure of DNA: The Double Helix
    Other DNA Structures
    Bends in DNA

II. IMPORTANT TERMS AND CONCEPTS

transforming principle	thymine
nuclease	uracil
DNase	deoxyribonucleotide
^{32}P and ^{35}S	ribonucleotide
tobacco mosaic virus	nucleoside
macromolecule	phosphodiester bond
nucleotide	polynucleotide
nitrogenous base	Chargaff's rules
ribose	X-ray diffraction
deoxyribose	antiparallel
purine	complementary base pair
pyrimidine	oligomer
adenine	A-DNA and B-DNA
guanine	Z-DNA
cytosine	*anti* and *syn* conformation

III. THINKING ANALYTICALLY

The realm of molecules is vast, but its units of measurement in relation to ordinary human experience are expressed for the most part by negative exponents. Take the distance, for example, separating the protons of the two hydrogen atoms in a molecule of hydrogen. It averages out somewhere around 0.74 Å (Angstrom), which in our world of intuitively comprehensible dimensions equals 0.74×10^{-10} meters, assuming of course that we are comfortable with meters. Perhaps we can get a better idea of how long an Angstrom is if we calculate how many of them there are in a meter. It comes out to 10,000,000,000, which would be a very nice number if it were your annual income in dollars. It wouldn't even be too hard to take in pennies.

    A couple of other units of measurement that will pop up when dealing with molecules are nanometers and, on rare occasion, daltons. One nanometer (1 nm) = 1×10^{-10} meter, and one dalton, which is a unit of mass, equals 1.650×10^{-24} grams.

A good piece of advice when working on a problem that presents values in a variety of units, is to convert those units to as small an assortment as possible. The following example will present an opportunity to do just that.

QUESTION EXAMPLE: If the average number of base pairs in a human chromosome is 111.7×10^6, and the average weight of a base pair, including the sugar and phosphate moieties associated therewith, is 660 daltons, and the distance between adjacent base pairs is 0.34 nanometers, answer the following:

How many feet of DNA are there in an average female human diploid cell, and how many grams does it weigh? (Hint: 1 meter (m) = 3.28 feet; the symbol # = "number"; chrom = chromosomes)

Solution:

    a. (i) [(# base prs)/(chrom)] x (# chrom's) = total # base prs
            $111.7 \times 10^6 \times 46 = 5138.2 \times 10^6 = 5.14 \times 10^9$
      (ii) (total # base prs) x (meters between base prs) = total m
            $5.14 \times 10^9 \times 0.34 \times 10^{-9} = 1.75$ m total length
      (iii) (total m length) x (ft/m) = total length in ft.
            $1.75 \times 3.28 = 5.74$ ft. (Answer)

    b. (i) (daltons / base pr) x (# base pr / chrom) = daltons/chrom
            $660 \times 111.7 \times 10^6 = 73722 \times 10^6$ daltons/chrom
      (ii) (daltons/chrom) x (# chroms) = total daltons
            $73.722 \times 10^9 \times 46 = 3391.2 \times 10^9$ total daltons
      (iii) (total dalton) x (grams/dalton) = total grams
            $3.39 \times 10^{12} \times 1.65 \times 10^{-24} = 5.59 \times 10^{-12}$ grams

Answer: With a mean diameter of approximately 2 nm, the total DNA in one diploid human cell would form a very fine thread about six feet long, but weighing only about 5.6×10^{-12} grams. This means it would be so fine as to be invisible.

IV. QUESTIONS FOR PRACTICE

A. Multiple Choice Questions

1.   DNA and RNA are polymers of
    a. nucleosides.
    b. nucleotides.
    c. pentose phosphates.
    d. ribonucleotides.

2.   The genetic material (genes) of eukaryotes is
    a. DNA.
    b. RNA.
    c. DNA and RNA.
    d. nucleoprotein.

3.   A molecule consisting of ribose covalently bonded to a purine or pyrimidine base is a
    a. ribonucleoside.
    b. ribonucleotide.
    c. nuclease.
    d. deoxyribonucleotide.

4.   The transforming principle was found to be
    a. acellular material that could alter a cell's heritable characteristics.
    b. a substance derived from killed viruses.
    c. modified RNA that could change a living cell.
    d. a transfusible substance that revives dead cells.

5. When Griffith injected mice with a mixture of live R pneumococcus that had been derived from a *IIS* strain and heat-killed *IIIS* bacteria,
   a. the mice survived and he recovered live type *IIIR* organisms.
   b. the mice died, but he recovered live type *IIIR* cells.
   c. the mice died, but he recovered live type *IIS*.
   d. the mice died, but he recovered live type *IIIS*.

6. Avery, MacLeod and McCarty demonstrated that the transforming principle was
   a. type *IIIS* protein.
   b. DNA.
   c. RNA.
   d. type *IIR* protein.

7. Analysis of the bases of a sample of nucleic acid yielded these percentages: A-20%, G-30%, C-20%, T-30%. Identify the sample as
   a. double-stranded RNA.
   b. double-stranded DNA.
   c. single-stranded RNA.
   d. single-stranded DNA.

8. Which of the chemical bonds below is the weakest?
   a. covalent
   b. phosphodiester
   c. hydrogen
   d. glycosidic

9. Rosalind Franklin deduced the helical structure of DNA from
   a. base pair analysis.
   b. X-ray diffraction.
   c. recombination data.
   d. melting points of DNA.

10. The arrangement of the two strands of DNA relative to each other is best described as
    a. random.
    b. unstable.
    c. parallel.
    d. antiparallel.

Answers: 1b, 2a, 3a, 4a, 5d, 6b, 7d, 8c, 9b, 10d

B. Thought Questions

1. Describe the various forms of DNA (A, B, Z) that have been identified and comment on their most significant differences. Have any functional differences been described or postulated, and if so, what are they? (cf. Text pp. 280-284).

2. Watson and Crick reputedly deduced the structure of DNA by applying observations made by others to molecular models, without any direct experimental observations of their own. Is this legitimate scientific procedure? If so, why? If not, why not?

3. Compare the experiments of Griffith; Avery, MacLeod and McCarty; and Hershey and Chase from the standpoint of rigorous application of scientific method and procedure. (cf. Text pp. 269-272)

V. ANSWERS AND SOLUTIONS TO TEXT QUESTIONS

9.1   In the 1920s while working with *Diplococcus pneumoniae*, the agent that causes pneumonia, Griffith discovered an interesting phenomenon. In the experiments mice were injected with different types of bacteria. For each of the following bacteria type(s) injected, indicate whether the mice lived or died:
    a. type *IIR;*
    b. type *IIIS;*
    c. heat-killed *IIIS;*
    d. type *IIR* + heat-killed *IIIS.*

Answer: a. lived
                b. died
                c. lived
                d. died (in this case DNA from the *S* bacteria transformed the *R* bacteria to a virulent form)

9.2   Several years after Griffith described the transforming principle, Avery, MacLeod, and McCarty investigated the same phenomenon.
    a. Describe their experiments.
    b. What did their experiments demonstrate beyond Griffith's?
    c. How were enzymes used as a control in their experiments?

Answer: a. They showed that transformation could occur in vitro, using extracts of *S* cells. The transforming principle copurified with DNA, indicating that the genetic material was DNA.
                b. The transforming principle was DNA.
                c. Ribonuclease did not destroy the transforming principle, but deoxyribonuclease did abolish transforming activity. This result substantiated the fact that DNA was the genetic material.

9.3   By differentially labeling the coat protein and the DNA of phage T2, Hershey and Chase demonstrated that (choose the correct answer)
    a. only the protein enters the infected cell.
    b. the entire virus enters the infected cell.
    c. a metaphase chromosome is composed of two chromatids, each containing a single DNA molecule.
    d. the phage genetic material is most probably DNA.
    e. the phage coat protein directs synthesis of new progeny phage.

Answer: d

9.4   Hershey and Chase showed that when phages were labeled with ^{32}P and ^{35}S, the ^{35}S remained outside the cell and could be removed without affecting the course of infection, whereas the ^{32}P entered the cell and could be recovered in progeny phages. What distribution of isotope would you expect to see if parental phages were labelled with isotopes of
    a. C?
    b. N?
    c. H?
    Explain your answer.

Answer: All three isotopes would appear both inside and outside the infected cell because C, N, and H occur in both DNA and proteins.

9.5   What is the evidence that the genetic material of TMV (tobacco mosaic virus) is RNA?

Answer: See text, p. 274.

9.6   In DNA and RNA, which carbon atoms of the sugar molecule are connected by a phosphodiester bond?

Answer: 3' and 5' carbons are connected by a phosphodiester bond.

9.7   Which base is unique to DNA and which base is unique to RNA?

Answer: Thymine is unique to DNA and uracil is unique to RNA. See text p. 277.

9.8   How do nucleosides and nucleotides differ?

Answer: A nucleoside in a nucleic acid is the sugar plus base, whereas a nucleotide is the sugar plus base plus phosphate. See text Fig. 9.12.

9.9   What chemical group is found at the 5' end of a DNA chain? At the 3' end of a DNA chain?

Answer: A phosphate group is found at the 5' end of a nucleic acid chain and a hydroxyl group is found at the 3' end.

9.10   What evidence do we have that in the helical form of the DNA molecule that the base pairs are composed of one purine and one pyrimidine?

Answer: When the chemical components of double-stranded DNA from a wide variety of organisms were analyzed quantitatively by Chargaff, it was found that the amount of purines equalled the amount of pyrimidines. More specifically, it was found the amount of adenine equalled the amount of thymine and the amount of guanine equalled the amount of cytosine. The simplest hypothesis was that complementary base pairing existed with A on one strand paired with T on the other strand, and similarly, G paired with C. More evidence came from X-ray diffraction analysis in which. the dimensions of the DNA double helix were established and compared with the known sizes of the bases. That is, the diameter of the double helix is constant throughout its length at 2 nm, which is the right size to accommodate a purine paired with a pyrimidine, but too small for a purine - purine pair, and too big for a pyrimidine - pyrimidine pair.

9.11   What evidence is there to substantiate the statement: "There are only two base pair combinations in DNA, A–T and C–G"?

Answer: The evidence for only two base pair combinations in DNA, that is A-T and G-C, comes from quantitative measurements of the four bases in double-stranded DNA isolated from a wide variety of organisms. In all cases, the amount of A was shown to equal the amount of T, and the amount of G was shown to equal the amount of C. Moreover, different DNAs exhibit different base ratios (stated as the %GC) so that while A = T and G = C, in most organisms (A+T) does not equal (G+C). The simplest hypothesis, therefore, is that there are two base pairs in DNA, A–T and G–C, and the proportion of the two base pairs varies from organism to organism.

9.12   How many different kinds of nucleotides are there in DNA molecules?

Answer: There are four different kinds of nucleotides in DNA molecules. A nucleotide is defined as base + sugar + phosphate. Each DNA nucleotide contains deoxyribose as the sugar and the four different nucleotides result from the presence of four different nitrogenous bases, adenine, thymine, guanine and cytosine. The four nucleotides are, therefore, deoxyadenosine monophosphate (dAMP), thymidine monophosphate (TMP), deoxyguanosine monophosphate (dGMP) and deoxycytidine monophosphate (dCMP).

9.13 What is the base sequence of the DNA strand that would be complementary to the following single-stranded DNA molecules:
   a. 5' AGTTACCTGATCGTA 3'
   b. 5' TTCTCAAGAATTCCA 3'

Answer: a. 3' TCAATGGACTAGCAT 5'
         b. 3' AAGAGTTCTTAAGGT 5'

9.14 Is an adenine-thymine or guanine-cytosine base pair harder to break apart? Explain your answer.

Answer: The adenine-thymine base pair is held together by two hydrogen bonds while the guanine-cytosine base pair is held together by three hydrogen bonds. Thus, the guanine-cytosine base pair requires more energy to break it and is the harder to break apart.

9.15 The double-helix model of DNA, as suggested by Watson and Crick, was based on a variety of lines of evidence gathered on DNA by other researchers. The facts fell into the following two general categories; give three examples of each:
   a. chemical composition;
   b. physical structure.

Answer: See text, pp. 278-280.

9.16 For double-stranded DNA, which of the following base ratios always equals 1?
   a. (A + T)/(G + C)
   b. (A + G)/(C + T)
   c. C/G
   d. (G + T)/(A + C)
   e. A/G

Answer: b, c, and d

9.17 If the ratio of (A + T) to (G + C) in a particular DNA is 1.00, does this result indicate that the DNA is most likely constituted of two complementary strands of DNA or a single strand of DNA, or is more information necessary?

Answer: More information is necessary since such a ratio says nothing about complementarily of bases. The bases in the numerator are not complementary to the bases in the denominator, and their ratio therefore may be equivalent to 1.00 or not equivalent to 1.00 in both single- and double-stranded DNA.

9.18 Explain whether the (A + T)/(G + C) ratio in double-stranded DNA is expected to be the same as the (A + C)/(G + T) ratio.

Answer: In double-stranded DNA the (A + C)/(G + T) ratio is expected to be equal to 1 because of the base pairing between A and T and between G and C. That is, A = T, G = C, and A + C = G + T; therefore (A + C)/(G + T) = 1. However. the (A + T)/(G + C) ratio may or may not be equal to 1 since there is no pairing between the bases in the numerator and the bases in the denominator. For example, a double-stranded molecule of DNA may have many A–T base pairs and few G–C base pairs.

9.19 The percent cytosine in a double-stranded DNA is 17. What is the percent of adenine in that DNA?

Answer: C = 17%, therefore G = 17% and the percent GC = 34. Therefore the percent AT = 66, and the percent A = 66/2 = 33.

9.20   Upon analysis, a DNA molecule was found to contain 32% thymine. What percent of this same molecule would be made up of cytosine?

Answer: In a double-stranded DNA molecule, A = T and G = C (the complementary base-pairing rule). Therefore, if the molecule contains 32% T (thymine) it must contain 32% A (adenine). Thus, the AT base pairs must constitute 64% of the molecule and the remaining 36% of the DNA must be GC base pairs. Since there is an equal amount of G and C in the DNA, 18% of the DNA would consist of cytosine.

9.21   A sample of double-stranded DNA has a percent GC content of 62. What is the percent of A in the DNA?

Answer: If the % GC = 62, then the % AT = 38. And since A = T in double-stranded DNA, the % A = 38/2 = 19.

9.22   A double-stranded DNA polynucleotide contains 80 thymidylic acid and 110 guanylic acid residues. What is the total nucleotide number in this DNA fragment?

Answer: The double-stranded DNA has 80 thymidylic acid residues and therefore must have 80 deoxyadenylic acid residues because of complementary base pairing rules. Similarly, the DNA has 110 deoxyguanylic acid residues and therefore must have 110 deoxycytidylic acid residues. The total number of nucleotides in the DNA fragment is 80 (T nucleotides) + 80 (A nucleotides) + 110 (G nucleotides) + 110 (C nucleotides) = 380 nucleotides, or 190 nucleotide pairs.

9.23   Analysis of DNA from a bacterial virus indicates that it contains 33% A, 26% T, 18% G, and 23 % C. Interpret these data.

Answer: If the DNA were double-stranded, we would expect complementary base-pairing. Since this is not the case, the DNA in question must be single-stranded.

9.24   What is a DNA oligomer?

Answer: A DNA oligomer is a short stretch of DNA containing only a few base pairs.

9.25   The genetic material of bacteriophage ΦX174 is single-stranded DNA. What base equalities or inequalities might we expect for single-stranded DNA?

Answer: We can make no predictions regarding the base content of single-stranded DNA. Any base-pair equality would depend on the overall sequence of the chromosome: A might be equal to T, but that result would be unlikely, given all the other possible sequences that might be the case.

9.26   If a virus particle contains double-stranded DNA with 200,000 base pairs, how many complete 360° turns occur in this molecule?

Answer: If there are 10 base pairs per complete 360° turn of double-stranded DNA, there are 200,000/10 = 20,000 complete turns in the viral DNA.

9.27   A double-stranded DNA molecule is 100,000 base pairs (100 kilobases long).
    a.   How many nucleotides does it contain?
    b.   How many complete turns are there in the molecule?
    c.   How long is the DNA molecule?

Answer: a.   Each base pair represents a pair of nucleotides, therefore 100,000 x 2 = 200,000 nucleotides.
    b.   There is a 3600 turn every 10 base pairs, therefore 100,000/ 10 = 10, 000 complete turns.
    c.   The vertical distance for each 360° turn = 3.4 nm. Therefore, since there are 10,000 complete turns, the vertical distance (i.e. length) is 3.4 nm x 10,000; i.e., 3.4×10^4 nm

9.28   If nucleotides were arranged at random in a single-stranded RNA 10^6 nucleotides long, and if the base composition of this RNA was 20% A, 25% C, 25% U and 30% G, how many times would you expect the specific sequence (5')-GUUA-(3') to occur?

Answer: The probability any group of four bases would be GUUA is (0.3)(0.25)(0.25)(0.2) = 0.00375. A molecule, 10^6 nucleotides long contains very nearly 10^6 groups of four bases. (The first group of four is bases 1, 2, 3 and 4, the second group of four is bases 2, 3, 4 and 5, etc.). Thus the number of occurrences of GUUA should be about (0.00375 x 10^6) = 3750.

CHAPTER 10

THE ORGANIZATION OF DNA IN CHROMOSOMES

I. CHAPTER OUTLINE

The Structural Characteristics of Prokaryotic and Viral Chromosomes
    Bacterial Chromosomes
    T-Even Phage Chromosomes
    Bacteriophage ΦX174 Chromosome
    Bacteriophage λ Chromosome
The Structural Characteristics of Eukaryotic Chromosomes
    The Eukaryotic Chromosome Complement
    The Karyotype
    Chromosomal Banding Patterns
    Cellular DNA Content and Phylogeny
    The Molecular Structure of the Eukaryotic Chromosome
    Centromeres and Telomeres
Sequence Complexity of Eukaryotic DNA
    Denaturation-Renaturation Analysis of DNA

II. IMPORTANT TERMS AND CONCEPTS

relaxed DNA	heterochromatin
supercoiled DNA	constitutive heterochromatin
positive & negative supercoiling	facultative heterochromatin
topoisomerase	nucleosome
DNA gyrase	nucleofilament
DNA domains	linker DNA
circularly permuted DNA	nucleosome core particle
exonuclease	30 nm chromatin fiber
concatamers	chromosome scaffold
headful packaging	centromere
cos sequence	telomere
Q banding	CEN sequence
G banding	telomere-associated sequences
C value	DNA denaturation and renaturation
C-value paradox	C_ot equation
histones	satellite DNA
nonhistone chromosome protein	unique sequence DNA
euchromatin	repetitive sequence DNA

III. THINKING ANALYTICALLY

Some of the problems concerning DNA and/or chromosomes require visualization of the presumed physical structure itself, in some cases in two dimensions, and in other cases in three. In these situations it is often helpful to make sketches that show the components in question. Text problem 10.1 is a case in point.

The following problem is quite different. Its solution depends upon simple analysis of the conditions put forth and recall of certain facts learned.

QUESTION EXAMPLE: The difference in denaturation time between two samples of two-stranded DNA is represented graphically below. The concentrations of the two samples were comparable, and the temperature was the same (96°C). The two samples differed in base pair ratio. The base pair ratio for Sample A was (A-T)/(G-C) = 4.0; for Sample B it was (A-T)/(G-C) = 0.25. How do you account for the difference in the time required for denaturation?

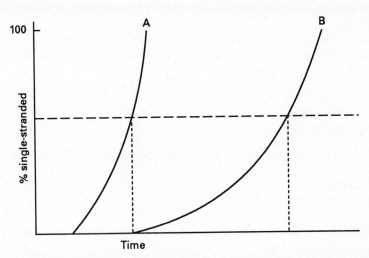

Solution: Denaturation of DNA by heat (melting) occurs as a consequence of weakening of the hydrogen bonds that hold the two strands together. Inspection of the molecular structure of the base pairs shows that the A-T pair is joined by two hydrogen bonds, and the G-C pair by three hydrogen bonds (See Fig. 9.16, Chapter 9). The base ratio of Sample A (4.0) means that 80% of the base pairs constitute 75%. The difference in denaturation times for the two samples, therefore, is most likely attributable to the greater number of hydrogen bonds that have to be broken in Sample B, owing to the preponderance of G-C base pairs.

IV. QUESTIONS FOR PRACTICE

A. Multiple Choice Questions

1.   The DNA of all known bacteria is
 a.   single stranded.
 b.   double stranded.
 c.   triple stranded.
 d.   either a or b, not c

2.   The chromosome of *E. coli* is packaged in the nucleoid region in a
 a.   nuclear membrane.
 b.   semicircular form.
 c.   relaxed form.
 d.   supercoiled form.

3.   The topoisomerases are enzymes that control the
 a.   base sequence of bacterial DNA.
 b.   rate of denaturation of DNA.
 c.   supercoiling of DNA in bacteria.
 d.   renaturation of DNA.

113

4. Topoisomerase II produces
   a. relaxed DNA.
   b. circular DNA.
   c. negative supercoiling.
   d. positive supercoiling.

5. Circularly permuted chromosomes characterize
   a. T2 and T4 phages.
   b. certain bacteria.
   c. all bacteria.
   d. all eukaryotes.

6. The DNA of the viruses ΦX174 and Qβ are
   a. single stranded.
   b double stranded.
   c. circularly permuted.
   d. extensively supercoiled.

7. The terminal redundancy that occurs in the chromosome of the T-even phages facilitates the formation of
   a. concatamers.
   b. circular permutations.
   c. neither a nor b
   d. both a and b

8. The total amount of DNA in the haploid genome of any organism is expressed as its
   a. C-value.
   b. C_ot equivalent.
   c. weight in daltons.
   d. length in centimeters.

9. An octamer of the histones H2A, H2B, H3 and H4 comprises a structure known as a
   a. telomere.
   b. centromere.
   c. nucleofilament.
   d. nucleosome particle.

10. The genomes of prokaryotes consist mostly of
    a. repetitive DNA sequences.
    b. interspersed DNA sequences.
    c. unique DNA sequences.
    d. clustered repeated DNA sequences.

Answers: 1b, 2d, 3c, 4a, 5a, 6a, 7d, 8a, 9d, 10c

B. Thought Questions

1. The histones, H1,H2A.H2B,H3 and H4, are said to be the most highly conserved of all proteins. From this fact, it is proposed that they serve a function that is basic to life for all eukaryotes. Comment on the implications of this assertion with regard to evolutionary theory. (See text p. 306ff.)

2. What do the nonhistone proteins include? How do they differ from the nonhistone chromosomal proteins? Name an example of each kind. (See text pp. 308 ff.)

3. How do you account for the presence of both unique and repetitive sequences in the genomes of eukaryotes? Why are the latter lacking in prokaryotes? (See text pp. 317-321.)

4. What are the different categories of chromatin, and in what ways do they differ both with respect to location and function? (See text pp. 307, 311-316.)

5. Describe the structural composition and organization of the chromosomes of eukaryotic cells. How do they compare with the chromosomes of viruses and of prokaryotes? (See text pp. 306-312.)

V. ANSWERS AND SOLUTIONS TO TEXT QUESTIONS

10.1 The nucleotide sequences of two DNA molecules from a population of T2 DNA molecules are as shown:

1 $\underrightarrow{\text{TAGCTCC}}$ 3 $\underrightarrow{\text{GCTCCTA}}$

2 $\underleftarrow{\text{ATCGAGG}}$ and   4 $\underleftarrow{\text{CGAGGAT}}$

These molecules were heat-denatured and then the separated strands were allowed to renature. Diagram the structures of the renatured molecules most likely to appear when (a) strand 2 renatures with strand 3 and (b) strand 3 renatures with strand 4. Mark the strands and indicate sequences and polarity.

Answer: a   You will notice that the sequence G C T C C in molecule 3 is complementary in the same order to the sequence C G A G G in molecule 2. Pair them up and see what it looks like:

$\underrightarrow{\text{G \quad C \quad T \quad C \quad C \quad T \quad A}}$

$\underleftarrow{\text{A \quad T \quad C \quad G \quad A \quad G \quad G}}$

You notice that each strand has two unpaired bases sticking out; but you also notice that they are complementary to each other. Knowing that the strands are flexible (after all, they do form supercoils, don't they?) try bending them around in a circle, so that the A's and T's base-pair. That's your answer!

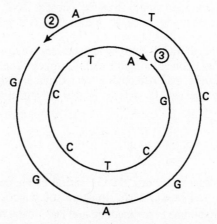

b   It is patently obvious that strands 3 and 4 are complementary, and the polarity is correct, so it is simply a matter of pairing them up:

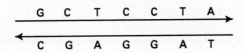

115

10.2   Capital letters represent regions in the chromosome of phage T4. A particular *E. coli* cell was infected by a single T4 chromosome with the sequence ABCDEFAB, but before this chromosome could replicate it suffered a deletion of the E region. This did not interfere with phage replication. What would you expect to be the chromosome sequence(s) of the progeny phage produced upon lysis of this cell? Explain your reasoning.

Answer: T4 chromosomes are circularly permuted and terminally redundant. Since progeny T4 chromosomes are cut by the "headful" from a long concatemer, the deletion of E will produce a shorter unique sequence, and result in longer terminal redundancy. Thus we should see ABCDFABC, BCDFABCD, CDFABCDF, DFABCDFA and FABCDFAB.

10.3
        a.   If you were to denature and renature a population of normal T4 chromosomes, what kinds of structures would form?
        b.   How would the results differ if the chromosomes were from T7?
        c.   From λ?

Answer: a.   The chromosomes are circularly permuted, terminally redundant molecules. Upon denaturation and renaturation, there would be few perfectly matched double-stranded molecules, but most molecules would have single-stranded ends. Molecules with very long single-stranded ends might form complex partially circularized structures.
        b.   All molecules would be perfectly matched.
        c.   λ chromosomes are linear; each chromosome has the same sequence and there are single-stranded "sticky ends" that permits the molecules to form circles. Therefore, all molecules would have single-stranded ends or be circularized (to nicked circles).

10.4   You are given a single-strand exonuclease (an enzyme that can digest single-stranded DNA from a free end but that cannot digest double-stranded DNA from a free end but that cannot digest double-stranded DNA). What would be the results of incubating each of the following phages with single-strand exonuclease and then using the phage to infect E. *coli?*
        a.   ΦX174.
        b.   λ.
        c.   T4.
        d.   T7.

Answer: a.   ΦX174 chromosomes are single-stranded, circular DNA molecules. ΦX174 would not be affected by this treatment because, although it is single stranded, it has no ends to be digested.
        b.   The single stranded "sticky" ends would be removed, which would prevent the λ phage from circularizing upon entering the cell. Lysogeny could not occur.
        c.   T4 would not be affected. Although it has redundant ends, they are not single stranded.
        d.   T7 would not be affected. Its ends are neither redundant nor single stranded.

10.5   What are topoisomerases?

Answer: See text p. 291.

10.6   In typical human fibroblasts in culture, the G1 period of the cell cycle lasts about 10 h, S lasts about 9 h, G2 takes 4 h, and M takes 1 h. Imagine you were to do an experiment in which you added radioactive (^{3}H) thymidine to the medium and left it there for 5 min (pulse), and then washed it out and put in ordinary medium (chase).
        a.   What % of cells would you expect to become labeled by incorporating the ^{3}H-thymidine into their DNA?
        b.   How long would you have to wait after removing the ^{3}H medium before you would see labeled metaphase chromosomes?
        c.   Would one or both chromatids be labeled?
        d.   How long would you have to wait if you wanted to see metaphase chromosomes containing ^{3}H in the regions of the chromosomes that replicated at the beginning of the S period?

Answer: a. The S phase is 9 h and the complete cell cycle takes 24 h. The proportion of the cell cycle in S is 9/24 = 0.375. Since S is 37.5% of the cycle, 37.5% of cells should be in S at any instant, and these are the ones that would incorporate label.

b. A little over 4 h. Such cells would have to take up some [3]H (a few minutes), pass through G2 (4 h), and pass through mitotic prophase (several minutes).

c. Both. Each chromatid is a double-stranded DNA molecule containing one old and one newly synthesized DNA strand.

d. A little more than 13 h. Such cells would have to pass through nearly all of S and all of G2 and through prophase.

10.7   Assume you did the experiment in Question 10.6, but left the radioactive medium on the cells for 16 h instead of 5 min. How would your answers to the above questions change?

Answer: a. All cells should be labeled. Those that had just barely finished S when you added label would have time to go through G2, M and G1 and get into S again and become labeled.

b. Labeled metaphases would already be present by the time the ^{3}H medium was removed.

c. Both

d. This would be difficult to do. If you waited a bit over 13 h you might catch a few cells that had just been labeled at the beginning of one S. You would also pick up cells that had been labeled at the end of one S as well as the beginning of the next.

10.8   Karyotype analysis performed on cells cultured from an amniotic fluid sample reveals that the cells contain 47 chromosomes. The Feulgen stained chromosomes are classified into groups and the arrangement shows 6 chromosomes in A, 4 in B, 16 in C, 6 in D, 6 in E, 4 in F, and 5 in the G group. Based on the above information
   a.   What could be the genotype of the fetus? If more than one possibility exists, give all.
   b.   How would you proceed to distinguish between the possibilities?

Answer: The fetus could be either a male Klinefelter syndrome XXY or an XX Down syndrome (with three #21 chromosomes), or, less likely, a female with three #22 chromosomes. The best way to distinguish between the possibilities is to do a Q banding. As is evident in Text Figure 10.14, the entire Y chromosome is intensely fluorescent and clearly diagnostic for the presence of Y. This stain also distinguishes chromosomes #21 from #22. The 3-21 would be a Down syndrome; 3-22 would not be a Down syndrome, but would be severely abnormal and likely to abort.

10.9   What is the relationship between cellular DNA content and the structural or organizational complexity of the organism?

Answer: There is no direct relationship. See text pp.303-304.

10.10   Match the DNA type with the chromatin type. (More than one DNA type may match a given chromatin type.)

DNA Type	Chromatin Type
Barr body (inactivated DNA)	Euchromatin
Centromere	Facultative heterochromatin
Telomere	Constitutive heterochromatin
Most expressed genes	

Answer: DNA type matched with chromatin type:

DNA Type	Chromatin Type
Barr body (inactivated DNA)	Facultative heterochromatin
Centromere	Constitutive heterochromatin
Telomere	Constitutive heterochromatin
Most expressed genes	Euchromatin

10.11 Eukaryotic chromosomes contain (choose the best answer)
    a. protein.
    b. DNA and protein.
    c. DNA, RNA, histone, and nonhistone protein.
    d. DNA, RNA, and histone.
    e. DNA and histone.

Answer: c.

10.12 List four.major features that distinguish eukaryotic chromosomes from prokaryotic chromosomes.

Answer: Here are five:
1. DNA of eukaryotic chromosomes is in linear form; prokaryotic "chromosome" is DNA in circular form,
2. Prokaryotic DNA associated with variety of proteins; eukaryotic DNA associated invariably with histones.
3. Eukaryotic chromosomes have centromeres and telomeres; prokaryotic chromosomes have neither,
4. Prokaryotic chromosomes have unique base sequences only; eukaryotic DNA has both unique and repetitive sequences,
5. A prokaryotic cell has only one chromosome; eukaryotic cells typically have multiple chromosomes.

10.13 Discuss the structure and role of nucleosomes.

Answer: See text pp. 308ff.

10.14 What are the main molecular features of yeast centromeres?

Answer: See text pp. 315-316.

10.15 What are telomeres?

Answer: See text pp. 316-317.

10.16 Would you expect to find most protein coding genes in unique-sequence DNA, in moderately repetitive DNA or in highly repetitive DNA?

Answer: Unique sequence. See text p. 320.

10.17 Would you expect to find tRNA genes in unique-sequence DNA, in moderately repetitive DNA, or in highly repetitive DNA?

Answer: In moderately repetitive DNA. See text p. 320.

10.18 The data plotted in the accompanying graph are obtained in a renaturation study of a DNA sample.

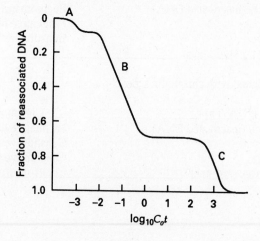

State whether each of the following is true or false. Explain your answer.
a. Satellite DNA is found in C fraction.
b. RNA and histone genes are found in B.
c. Telomere and centromere sequences are found in A.
d. DNA coding for mRNA is found in A.
e. The source of the DNA used in this experiment could be eukaryotic or prokaryotic.

Answer:
    a. False. Satellite DNA would have highly repetitive sequences and would be found in the A fraction.
    b. True. These are moderately repetitive sequences.
    c. True. These are highly repetitive sequences.
    d. False. DNA coding for mRNA would be in the unique sequence fraction.
    e. False. Prokaryotic DNA is mostly unique sequence. No measurable amounts of repetitive sequences can be detected by C_ot analysis. The DNA source must have been a eukaryote.

10.19   A particular virus has a genome consisting of 10^5 bp of double stranded DNA. When this DNA is denatured and renatured, it reaches 50% renaturation at a C_ot of 3×10^{-1}, and a monophasic renaturation curve is seen. In contrast, when the DNA of a particular amphibian is denatured and renatured, a biphasic renaturation curve is seen. About 50% of this DNA reaches 50% renaturation at a C_ot of 3, while the other half of the DNA reaches 50% renaturation at a C_ot of 3×10^4.
a. What is the size of the amphibian genome?
b. What is the significance of the biphasic renaturation curve shown by the amphibian DNA?
c. How many copies of the C_ot 3 species are present in the amphibian genome?
d. How big are the repeats?

Answer:
    a. Assuming the virus and the C_ot 3×10^4 amphibian species both represent unique sequences, the amphibian genome is 5×10^9 bp. There is a 10^5 fold difference in between the C_ot two species, but the unique sequences are only half of the amphibian genome.
    b The two phases of the curve result from the renaturation of repeated (C_ot 3) and unique (C_ot 3×10^4) sequences.
    c This population has a C_ot 10^4 times lower than the unique sequences, so it must be 10^4 fold more abundant. Thus there are 10^4 copies per genome.
    d The 10^4 copies represent half of the 5×10^9 genome, or 2.5×10^9 bp. Each copy is thus $(2.5 \times 10^9)/10^4$ or 2.5×10^5 bp long.

CHAPTER 11

DNA REPLICATION AND RECOMBINATION

I. CHAPTER OUTLINE

DNA Replication in Prokaryotes
    Early Models for DNA Replication
    The Meselson-Stahl Experiment
    DNA Synthesis Enzymes
    Molecular Details of DNA Replication
    Bacterial DNA Replication and the Cell Cycle
DNA Replication in Eukaryotes
    Molecular Details of DNA Synthesis in Eukaryotes
    Assembly of New DNA into Nucleosomes
    Genetics of the Eukaryotic Cell Cycle
DNA Recombination
    Crossing-Over: Breakage and Rejoining of DNA
    The Holliday Model for Recombination
    Mismatch Repair and Gene Conversion

II. IMPORTANT TERMS AND CONCEPTS

semiconservative model
conservative model
dispersive model
density gradient centrifugation
DNA polymerase
dNMP
enzyme active site
enzyme substrate
enzyme product
conditional mutant
exonuclease
semidiscontinuous DNA replication
replication fork
replication bubble

DNA helicase
DNA primase
DNA ligase
Okazaki fragments
bidirectional replication
replisome
rolling circle replication
single-strand DNA binding proteins
Holliday model
Holiday intermediate
patched duplex
spliced duplex
heteroduplex
gene conversion

III. ANALYTICAL THINKING

The sort of material covered in this chapter requires very close attention to details. It is recommended that you make frequent reference to the figures and diagrams as you read the text, point-by-point, making sure that you have a clear comprehension of each detail before going on to the next. Some students find it helpful to close the book and try to reproduce the more complex diagrams on scratch paper. In fact, you might find that they can be improved upon!

IV. QUESTIONS FOR PRACTICE

A. Multiple Choice

1. Which of the following models for DNA replication is supported best by experimental evidence?
   a. semiconservative
   b. dispersive
   c. conservative
   d. semidispersive

2. The enzymes that are most directly concerned with catalyzing the synthesis of DNA are the DNA
   a. ligases.
   b. exonucleases.
   c. polymerases.
   d. primases.

3. Which of the following is *not* essential for the in vitro synthesis of DNA?
   a. magnesium ions
   b. DNA polymerase
   c. DNA primase
   d. a DNA fragment

4. DNA primase is a derivative of
   a. RNA polymerase.
   b. DNA ligase.
   c. DNA exonuclease.
   d. neither a, b, nor c

5. A critical part of any enzyme is
   a. a terminal phosphate.
   b. a phosphodiester bond.
   c. a suitable substrate.
   d. its active site.

6. As the parent strands of DNA separate during replication, they must untwist. The untwisting process is catalyzed by the enzyme
   a. DNA primase.
   b. DNA helicase.
   c. DNA replicase.
   d. DNA ligase.

7. As the parent strands separate during DNA replication, any tendency they might have to undergo intramolecular base pairing is counteracted by
   a. rolling circle model.
   b. single-strand DNA-binding proteins (SSB).
   c. autonomously replicating sequences (ARS).
   d. the primase-helicase complex.

8. DNA replication in certain viruses, such as ΦX174, and in *E. coli* F-factor during conjugation is executed by the so-called
   a. rolling circle model.
   b. semidiscontinuous model.
   c. DNA fragmentation model.
   d. consensus sequence.

9. The first stage of genetic recombination according to the Holliday model is
   a. the first gap phase (G_1).
   b. chromosome condensation.
   c. spindle fiber attachment.
   d. recognition and alignment.

10. Mismatch repair of heteroduplex DNA can result in
    a. restoration of allelic segregation.
    b. nullification of allelic segregation.
    c. a non-Mendelian segregation of alleles.
    d. genetic diversity.

Answers: 1a, 2c, 3c, 4a, 5d, 6b, 7b, 8a, 9d, 10c

B.   Thought Questions

1. In his investigations of the requirements for *in vitro* synthesis of DNA, Kornberg found that an absolute minimum of four components was necessary. What were these components and why in particular were each and all necessary? Also, what limitations did his *in vitro* method have as compared to in *vivo* synthesis, which involves other components, such as DNA helicase, DNA primase, and others? (Hint: See text pp. 329-333, 334, 341.)
2. Describe the mechanism of semidiscontinuous DNA replication. See text pp. 334-339.
3. Describe how the assembly of eukaryote DNA into nucleosomes is believed to be accomplished. See text pp. 349-350.
4. Describe the phenomena of mismatched repair and gene conversion. (See text p. 355.) What evolutionary significance might they imply, if any?

V.   ANSWERS AND SOLUTIONS TO TEXT QUESTIONS

11.1   Compare and contrast the conservative and semiconservative models for DNA replication.

Answer: In the semiconservative model the two strands of the double helix separate and each serves as a template for new DNA synthesis. Each daughter DNA double helix has one old and one new DNA strand. In the conservative model the two strands remain together to serve as the template for new DNA synthesis. One daughter helix consists of both parental strands, and the other consists of two new strands.

11.2   Describe the Meselson and Stahl experiment, and explain how it showed that DNA replication is semiconservative.

Answer: See text pp. 327-329. The key result in the experiment was the presence in the first generation of only DNA with intermediate density.

11.3   In the Meselson and Stahl experiment ^{15}N-labeled cells were shifted to ^{14}N medium, at what we can designate as generation 0.
   a. For the semiconservative model of replication, what proportion of ^{15}N-^{15}N, ^{15}N-^{14}N, and ^{14}N-^{14}N would you expect to find at generations 1, 2, 3, 4, 6, and 8?
   b. Answer the above question in terms of the conservative model of DNA replication.

Answer: Key: ^{15}N-^{15}N DNA = HH; ^{15}N-^{14}N DNA = HL; ^{14}N-^{14}N DNA = LL.
   a. Generation 1: all HL; 2: 1/2 HL, 1/2 LL; 3: 1/4 HL, 3/4 LL; 4: 1/8 HL, 7/8 LL; 6: 1/32 HL, 31/32 LL; 8: 1/128 HL, 127/128 LL
   b. Generation 1: 1/2 HH, 1/2 LL; 2: 1/4 HH, 3/4 LL; 3: 1/8 HH, 7/8 LL; 4: 1/16 HH, 15/16 LL; 6: 1/64 HH, 63/64 LL; 8: 1/256 HH, 255/256 LL.

11.4 Suppose *E. coli* cells are grown on an ^{15}N medium for many generations. Then they are quickly shifted to an ^{14}N medium, and DNA is extracted from the samples taken after one, two, and three generations. The extracted DNA is subjected to equilibrium density gradient centrifugation in CsCl. In the figure below, using the reference positions of pure ^{15}N and pure ^{14}N DNA as guides, indicate where the bands of DNA would equilibrate if replication were semiconservative or conservative.

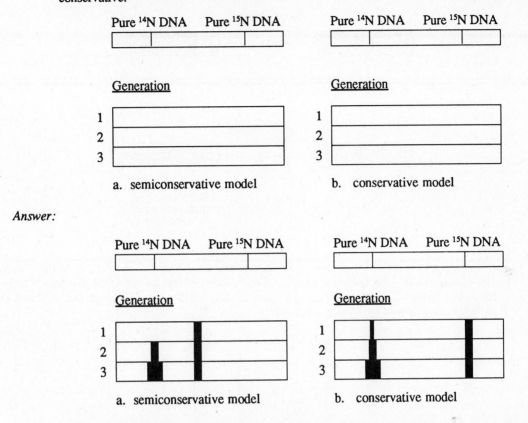

Answer:

11.5 A spaceship lands on earth and with it a sample of extraterrestrial bacteria. You are assigned the task of determining the mechanism of DNA replication in this organism.

You grow the bacteria in unlabeled medium for several generations, then grow it in presence of ^{15}N exactly for one generation. You extract the DNA and subject it to CsCl centrifugation. The banding pattern you find is shown in the figure below.

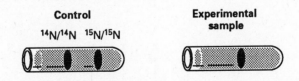

It appears to you that this is evidence that DNA replicates in the semiconservative manner, but this result does not prove that this is so. Why? What other experiment could you perform (using the same sample and technique of CsCl centrifugation) that would further distinguish between semiconservative and dispersive modes of replication?

123

Answer: The CsCl result eliminates the possibility of the conservative model of replication, but is consistent with either semiconservative or dispersive models. To distinguish between these two possibilities, using the same DNA sample isolated after one generation in ^{15}N, one could denature that DNA and then subject the single stranded sample to CsCl centrifugation:

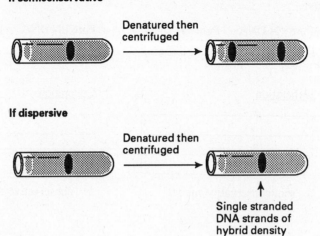

11.6 Assume you have a DNA molecule with the base sequence T-A-T-C-A going from the 5' to the 3' end of one of the polynucleotide chains. The building blocks of the DNA are drawn as in the following figure. Use this shorthand system to diagram the completed double stranded DNA molecule, as proposed by Watson and Crick.

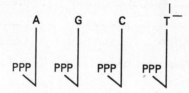

Answer:

a. You know that the triphosphates are bonded to the number 5 carbon of the deoxyribose molecule, which establishes the 5' end of the deoxyribonucleotides (dATP, dGTP, dTTP, and dCTP).

b. You know that the phosphodiester bond will form between this 5' end and the number 3 carbon of the next deoxyribonucleotide, replacing its OH group. Consequently the 5' to 3' bonding can be represented for the T-A-T-C-A sequence as:

c. Then by base pairing with antiparallel orientation, you get:

11.7 List the components necessary to make DNA *in vitro* by using the enzyme system isolated by Kornberg.

Answer: See text, pp. 328-333. DNA polymerase; intact, high-molecular-weight DNA; dATP, dGTP, dTTP, and dCTP, and magnesium ions are needed.

11.8 Give two lines of evidence that the Kornberg enzyme is not the enzyme involved in the replication of DNA for the duplication of chromosomes in growth of *E. coli*.

Answer: See text, pp. 330-333.

11.9 Base analogs are compounds, that resemble the natural bases found in DNA and RNA, but are not normally found in those macromolecules. Base analogs can replace their normal counterparts in DNA during *in vitro* DNA synthesis. Four base analogs were studied for their effects on *in vitro* DNA synthesis using the *E. coli* DNA polymerase. The results were as follows, with the amounts of DNA synthesized "pressed as percentages of that synthesized from normal bases only.

Normal Bases Substituted
By The Analog

ANALOG	A	T	C	G
A	0	0	0	25
B	0	54	0	0
C	0	0	100	0
D	0	97	0	0

Which bases are analogs of adenine? of thymine? of cytosine? of guanine?

Answer: None is an analog of adenine.
B and D are analogs of thymine.
C is an analog of cytosine.
A is an analog of guanine.

11.10 Describe the semidiscontinuous model for DNA replication. What is the evidence showing that DNA synthesis is discontinuous on at least one template strand?

Answer: See text, pp. 334-339.

11.11 Distinguish between a primer strand and a template strand.

Answer: A primer strand is a nucleic acid sequence that is extended by DNA synthesis activities. A template strand directs the base sequence of the DNA strand being made; for example, an A on the template causes a T to be inserted on the new chain, and so on.

11.12   The length of the *E. coli* chromosome is about 1100 μm.
    a.   How many base pairs does the *E. coli* chromosome have?
    b.   How many complete turns of the helix does this chromosome have?
    c.   If this chromosome replicated unidirectionally and if it completed one round of replication in 60 min, how many revolutions per minute would the chromosome be turning during the replication process?
    d.   The *E. coli* chromosome, like many others, replicates bidirectionally. Draw a simple diagram of a replicating *E. coli* chromosome that is halfway through the round of replication. Be sure to distinguish new and old DNA strands.

Answer: a.   One base pair is 0.34 nm, and the chromosome is 1100 μm. So the number of base pairs is $(1100/0.34) \times 1000 = 3.24 \times 10^6$.
    b.   There are 10 base pairs per turn in a normal DNA double helix; therefore it has a total of 3.24×10^5 turns.
    c. 3.24×10^5 turns and 60 min for unidirectional synthesis; therefore $(3.24 \times 10^5/60)$ turns per minute = 5400 revolutions per minute.
    d.

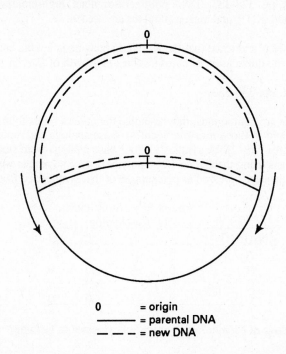

0   = origin
———— = parental DNA
– – – – = new DNA

11.13   In *E. coli* the replication fork moves forward at 500 nucleotide pairs per second. How fast is the DNA ahead of the replication fork rotating?

Answer: There are ten base pairs per helical turn, so 500 nucleotide pairs corresponds to the helix rotating 50 times per second, or 300 revolutions per minute.

11.14   A diploid organism has 4.5×10^8 base pairs in its DNA. This DNA is replicated in 3 minutes. Assuming all replication forks move at a rate of 10^4 base pairs per minute, how many replicons (replication units) are present in this organism's genome?

Answer: There are 4.5×10^8 of DNA and all is replicated in 3 minutes at 10^4 bp/min. Therefore, there are $(4.5 \times 10^8/(3 \times 10^4 \times 10^4))$ replication forks. Since it is a diploid organism, there are $(1.5 \times 10^4)/2 = 7.5 \times 10^3$ replicons in the genome.

11.15   The following events, steps or reactions occur during *E. coli* DNA replication. For each entry in Column A, select the appropriate entry in Column B. Each entry in A may have more than one answer, and each entry in B can be used more than once.

Column A		Column B	
_____a.	Unwinds the double helix	A	Polymerase I
		B	Polymerase III
		C	Helicase
_____b.	Prevents reassociation of complementary bases	D	Primase
		E	Ligase
		F	SSB protein
_____c.	Is a RNA polymerase	G	Gyrase
		H	None of these
_____d.	Is a DNA polymerase		
_____e.	Is the "repair" enzyme		
_____f.	Is the major elongation enzyme		
_____g.	A 5'-3' polymerase		
_____h.	A 3'-5' polymerase		
_____i.	Has 5'-3' exonuclease function		
_____j.	Has 3'-5' exonuclease function		
_____k.	Bonds free 3'-OH end of a polynucleotide to a free 5' monophosphate end of polynucleotide		
_____l.	Bonds 3'-OH end of a polynucleotide to a free 5' nucleotide triphosphate		
_____m.	Separates daughter molecules and causes supercoiling		

Answer: a. C
             b   F
             c   D
             d. A, B
             e   A
             f. B
             g. A, B, D
             h. H
             i. A, B
             j. A, B
             k. E
             l. D
             m. G

11.16   Compare and contrast the three *E. coli* DNA polymerases with respect to their enzymatic activities.

Answer: See text, pp. 334-339; see Table 11.1; they differ in associated exonuclease activities.

11.17   In *E. coli,* distinguish between the activities of primase; single-stranded, binding protein; helicase; DNA ligase; DNA polymerase I; and DNA polymerase III in DNA replication.

Answer: See text, pp. 337-339.

11.18   Describe the molecular action of the enzyme DNA ligase. What properties would you expect an *E.coli* cell to have if it had a temperature-sensitive mutation in the gene for DNA ligase?

Answer: DNA ligase seals single-stranded gaps in a DNA double helix in which there is a 3'-OH and a 5'-monophosphate. The enzyme is used to join Okazaki fragments produced by the discontinuous mechanism of DNA replication, so a temperature sensitive DNA ligase mutant would not be able to join these fragments at high temperature. The accumulation of such fragments in a temperature-sensitive ligase mutant strain grown at the nonpermissive temperature was used as evidence for the discontinuous model of replication-

11.19   Chromosome replication in *E. coli* commences from a constant point, called the origin of replication. The results of autoradiography experiments suggested to Cairns that chromosome replication was unidirectional. It is now known that DNA replication is bidirectional. Devise a biochemical experiment to prove that the *E. coli* chromosome replicates bidirectionally. (Hint: Assume that the amount of gene product is directly proportional to the number of genes.)

Answer: Develop a way to assay for gene products of genes on either side of origin and for various other points around the chromosomes If replication is bidirectional, there should be a doubling of the gene products both clockwise and counterclockwise from the origin. The wave of replication in both directions can be followed in time by such assays.

11.20   What property of DNA replication was indicated by the presence of Okazaki fragments?

Answer: Okazaki fragments indicate that DNA replication occurs in a discontinuous fashion. We know that this occurs because the replication fork moves in one direction, while the two DNA strands are of opposite polarity, necessitating continuous DNA synthesis on the leading strand template and discontinuous DNA synthesis on the lagging strand template.

11.21   A space probe returns from Jupiter and brings with it a new microorganism for study. It has double-stranded DNA as its genetic material. However, studies of replication of the alien DNA reveal that, while the process is semi-conservative, DNA synthesis is continuous an both the leading and the lagging strand templates. What conclusion(s) can you make from that result?

Answer.   The first conclusion one can make is that DNA replication in the Jovian bug does not occur as it does in *E. coli.* Two hypotheses can be proposed: (1) The Jovian microorganism contains a DNA polymerase that can synthesize DNA in the 3'-to-5' direction as well as one that can synthesize DNA in the 5'-to-5' direction; (2) The double-stranded DNA is not organized in an-anti parallel way as is *E. coli* DNA, so that DNA polymerases can synthesize new DNA in the 5'-to-3' direction on both template strands as the replication fork moves.

11.22   Compare and contrast eukaryotic and prokaryotic DNA polymerases.

Answer: See text Tables 11.1 and 11.4.

11.23   Draw a eukaryotic chromosome as it would appear at each of the following cell cycle stages. Show both DNA strands, and use different line styles for old and newly synthesized DNA.
   a. G_1
   b.   anaphase of mitosis
   c. G_2
   d.   anaphase of meiosis I
   e.   anaphase of meiosis II

Answer:

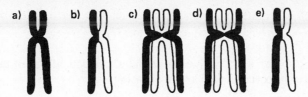

a) b) c) d) e)

11.24 Autoradiography is a technique which allows radioactive areas of chromosomes to be observed under the microscope. The slide is covered with a photographic emulsion, which is exposed by radioactive decay. In regions of exposure the emulsion forms silver grains upon being developed. The tiny silver grains can be seen on top of the (much larger) chromosomes. Devise a method for finding out which regions in the human karyotype replicate during the last 30 min of the S period. (Assume a cell cycle in which the cell spends 10 h in G_1, 9 h in S, 4 h in G_2 and 1 h in M.)

Answer: Assuming these cells have the typical 4 h G_2 period, you would add ^3H thymidine to the medium, wait 4.5 h, and prepare a slide of metaphase chromosomes. Autoradiography would then be done on this chromosome preparation. Regions of chromosomes displaying silver grains are then the late-replicating regions. Cells which were at earlier stages in the S period when they began, to take up ^3H will not have had sufficient time to reach metaphase.

11.25 When the eukaryotic chromosome duplicates, the nucleosome structures must duplicate. Discuss the synthesis of histones in the cell cycle, and discuss the model for the assembly of new nucleosomes at the replication forks.

Answer: See text pp. 349-350.

11.26 How is mismatch repair related to recombination and repair of DNA?

Answer: See text, pp. 352-354. Hybrid DNA resulting from branch migration may contain mismatched sequences. Mismatched sequences are recognized by special enzymes that remove a short segment of one DNA strand and then fill in the gap. The mismatch repair does not recognize which is the correct sequence.

11.27 Crosses were made between strains, each of which carried one of three different alleles of the same gene, *a*, in yeast. For each cross, some unusual tetrads resulted at low frequencies. Explain the origin of each of these tetrads:

Cross: a1 a2$^+$ a1 a3$^+$ a2 a3$^+$
         x   x   x
    a1$^+$a2   a1$^+$a3   a2$^+$a3

Tetrads: a1$^+$ a2 a1$^+$ a3 a2$^+$ a3
        a1$^+$ a2$^+$ a1$^+$ a3 a2$^+$ a3$^+$
        a1 a2 a1$^+$ a3$^+$ a2 a3$^+$
        a1 a2$^+$ a1 a3$^+$ a2 a3$^+$

Answer: In each row there is evidence that the segregation of one of the alleles in the tetrad has resulted from gene conversion caused by mismatch repair of heteroduplex DNA.
For the *a1 a2$^+$* x *a1$^+$ a2* cross the tetrad shows 2:2 segregation of *a1$^+$:a1* but 3:1 segregation of *a2$^+$:a2*, indicating gene conversion of an *a2* allele to its wild-type counterpart. Similarly, the *a1 a3$^+$* x *a1$^+$ a3* cross shows 2:2 segregation of *a3$^+$:a3* and 3:1 segregation of *a1$^+$:a1* resulting from gene conversion of an *a1* allele to *a1$^+$*. In the *a2 a3$^+$* x *a2$^+$ a3* cross,

the *a2* allele segregates in a Mendelian fashion while the *a3* allele segregates 3:1 a3$^+$:a3 as a result of gene conversion of one *a3* allele to *a3*$^+$.

11.28   From a cross of *y1 y2*$^+$ x *y1*$^+$ *y2*, where *y1* and *y2* are both alleles of the same gene in yeast, the following tetrad type occurs at very low frequencies:

$$
\begin{array}{ll}
y1^+ & y2 \\
y1 & y2 \\
y1 & y2 \\
y1 & y2^+
\end{array}
$$

Explain the origin of this tetrad at the molecular level.

Answer: The tetrad shows evidence of gene conversion of both alleles: a wild-type allele of *y1* has undergone gene conversion to a y1 allele, and a wild-type allele of *y2* has undergone gene conversion to a *y2* allele.

11.29   In *Neurospora the a, b,* and *c* loci are situated in the same arm of a particular chromosome. *a* is near the centromere, *b* is near the middle, and *c* is near the telomere of the arm. Among the asci resulting from a cross of *ABC* x *abc,* the following ascus was found (the 8 spores are indicated in the order in which they were arranged in the ascus): *ABC, ABC, ABc, ABc, aBC, aBC, abc, abc.* How might this ascus have arisen?

Answer: An event of recombination has occurred between *a* and *c* (*c* shows second division segregation, *a* does not). Evidently the exchange took place in the vicinity of *b*. The figure below shows a recombinational intermediate and its resolution in such a way that the two recombinant chromatids contain base mismatches. Repair of the mismatches happened to occur in the same direction in both chromatids, producing the 3*B*:1*b* segregation.

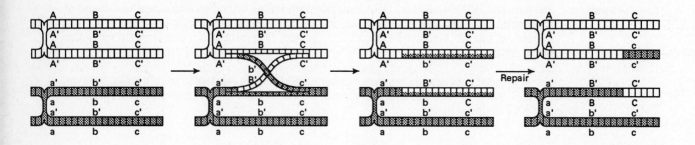

11.30   In the population of asci produced in question 11.29 an ascus was found containing, in this order, the spores *ABC, ABC, ABc, Abc, aBC, aBC, abc, abc.* How could this ascus have arisen?

Answer: The same series of events as in question 11.29 occurred, except that mismatch repair did not take place in one of the recombinant chromatids. DNA replication and the ensuing mitotic division that turned the four products of meiosis into eight spores caused the mismatched strands to be segregated into spores 3 and 4.

CHAPTER 12

TRANSCRIPTION

I. CHAPTER OUTLINE

Overview of Transcription
Basic Properties of Transcription
Transcription of Bacterial and Bacteriophage Genes
    RNA Polymerase in Bacteria
    Initiation and Elongation
    Termination of Transcription
    RNA Polymerases and Phage Transcription
Transcription in Eukaryotes
    RNA Polymerase in Eukaryotes
    Transcription of the Different Classes of Eukaryotic Genes

II. IMPORTANT TERMS AND CONCEPTS

transcription & translation
m, t, r, and snRNA
structural gene
ribonuclease
RNA polymerase
template & non-template strand
core enzyme *vs.* sigma
rpo genes
promoter
holoenzyme
terminator
consensus sequence
Boxes- Pribnow, TATA, CAAT, GC,
        A, B, C

rho
hairpin loop
transcription factors
regulatory factors
enhancer *vs.* silencer
promoter element
UAS
initiation complex
internal control region
transcription unit
spacer &, NTS sequence
termination site
core promoter

III. THINKING ANALYTICALLY

The material in Chapter 12 is primarily descriptive and the task of the student at the introductory level will be to organize the information in such a way that it will be remembered.

As a beginning, it will help to go through the list of terms and concepts with a view toward identifying the chemical nature of each of the items. Is the item a region of DNA, RNA, or does it refer to all or part of a protein? It may help to ask yourself the question "Does it do something?", because, if the answer is "yes," the item is probably a protein or polypeptide.

Next, distinguish between processes and characteristics of prokaryotes and eukaryotes. The fact that three RNA polymerases are active in eukaryotes results in three sets of data regarding initiation and termination, including sites where each occur and the factors that are essential in each case. It is often helpful to prepare a list of analogous sites and factors, pairing those elements that do similar jobs in prokaryotes with those in eukaryotes.

Finally, the matters of symbolic convention need to be explicitly learned. Understand which strand is conventionally indicated, and what this means as DNA is transcribed in the 3' to 5' direction.

Some of the practice problems that follow will help you organize the data.

IV. QUESTIONS FOR PRACTICE

A. Multiple Choice Questions

1. Match the item in Column A with one of the items in the answer list.

Column A		Answer List	
1.	Sigma	a	DNA
2.	Promoter	b.	RNA
3.	Terminator	c.	Protein
4.	Transcription factors	d.	More than one of the above.
5.	Regulatory factors			
6.	Pribnow Box			
7.	Transcription initiation, complex			
8.	Spacer sequences			
9.	NTS sequence			
10.	Hairpin loop			

Answer: 1c, 2a, 3a, 4c, 5c, 6a, 7d, 8a, 9a, 10a

2.   In most organisms, the DNA that is transcribed
   a.   is wholly located on the template strand.
   b.   is wholly located on the non-template strand.
   c.   is divided between the template and nontemplate strands.
   d.   is copied in the 5' to 3' direction.

3.   RNA polymerase differs from DNA polymerase in that
   a.   RNA polymerase is made of RNA whereas DNA polymerase is made of DNA.
   b.   only DNA polymerase can proofread.
   c.   only RNA polymerase can initiate the formation of a polynucleotide.
   d.   more than one of the preceding is correct

4. Which of the following recognizes and binds to a promoter?
   a.   The Pribnow box.
   b.   Sigma.
   c.   Omega.
   d.   Core enzyme.

5.   The RNA polymerases found in mitochondria and chloroplasts
   a.   are the same as those of their host cell.
   b.   are identical to bacteriophage RNA polymerase.
   c.   are identical to bacterial RNA polymerase.
   d.   are uniquely characteristic of those organelles.

Answer: 2c, 3d, 4b, 5d

B. Thought Questions

1. Compare and contrast each of the following pairs of terms with respect to location, function, and host organism (prokaryote or eukaryote).
   a. Pribnow Box & TATA Element. (See text pp. 367-368 and 373-375).
   b. Promoter elements, enhancer elements, silencer elements, and UAS (see text pp.373-374).
   c. Transcription in T7 and T4 bacteriophages. (See text p. 371.)
   d. The promoters that characterize RNA polymerase I, II, and III.
   e. Promoters that are internally located within the gene that is transcribed and upstream promoters. (See text pp. 372-376.)
2. Do prokaryotes or eukaryotes have more RNA polymerase (per unit total protein? Why should the two groups of organisms differ? (See text p. 372.)
3. How does *rho*-dependent termination differ from *rho*-independent termination, and, what is the role of the hairpin loop in each of these events? (see text p. 370-371)
4. The promoters for polymerase II have been highly conserved and, thus, differ little among eukaryotes. However, the promoters for polymerase I show significant differences among eukaryotes. Can you think of a logical reason for this difference? (Hint: How many genes are read by each these polymerases?)
5. What takes the place of sigma in eukaryotes? (See text p.367 for a discussion of transcription initiation factors.)

V. ANSWERS AND SOLUTIONS TO TEXT QUESTIONS

12.1 Describe the differences between DNA and RNA.

Answer: The DNA contains deoxyribose and thymine, whereas RNA contains ribose and uracil, respectively. Also, DNA is usually double-stranded, while RNA is usually single-stranded.

12.2 Compare and contrast DNA polymerases and RNA polymerases.

Answer: See text for this chapter and for Chapter 11, Tables 11.1 and 11.4, and Chapter 12, Table 12.1. Both DNA polymerases and RNA polymerases catalyze the synthesis of the nucleic acids in the 5'-to-3' direction. Both must recognize DNA and ensure the correct base pairing during the synthesis process. The RNA polymerases usually recognize specific base-pair sequences as signals for where to start transcription, which is generally not the, case for DNA polymerases, The DNA polymerases cannot initiate a DNA chain, whereas RNA polymerases can.

12.3 On page 363 you will find the sentence "All base pairs in the genome are replicated during the DNA synthesis phase of the cell cycle, but only *some* of the base pairs are transcribed into RNA." How is it determined *which* base pairs of the genome are transcribed into RNA?

Answer: The DNA base sequence determines this. Regions not bounded by transcription-initiation and transcription-termination signals are not transcribed. Furthermore, in regions bounded by these signals, only one strand is ordinarily transcribed. In addition, regions bounded by these signals may only be transcribed if certain transcription-inducing molecules are present.

12.4 Discuss the structure and function of the *E. coli* RNA polymerase. In your answer, be sure to distinguish between RNA core polymerase and RNA core polymerase-sigma factor complex.

Answer: See text pp. 365-371.

12.5 Discuss the similarities and differences between the *E. coli* RNA polymerase and eukaryotic RNA polymerases.

Answer: See text pp. 363-369, 371-373.

12.6 The RNA polymerases bind to promoter sequences in the DNA.
    a. Compare and contrast promoter sequences from prokaryotes and eukaryotes.
    b. Apart from the sequence information available for promoters for a number of genes, what evidence is there that promoters do not all have the same base sequence?

Answer: See text pp. 367-368, 372-373, 377, and Figures 12.8 and 12.11.

12.7 Discuss the molecular events involved in the termination of RNA transcription in prokaryotes.

Answer: See text pp. 370-371. Specific base-pair sequences are involved. In prokaryotes, transcription termination involves a complex between the RNA polymerase and the *nus* A protein which interacts with the termination sequence. The protein factor *rho* is also involved with transcription termination; it plays a role in RNA transcript release from the DNA.

12.8 Which classes of RNA do each of the three eukaryotic RNA polymerases synthesize? What are the functions of the different RNA types in the cell?

Answer: RNA polymerase I transcribes the major rRNA genes that code for 18S, 5.8S and 28S rRNAs; RNA polymerase II transcribes the protein-coding genes to produce mRNA molecules; and RNA polymerase III transcribes the 5S rRNA genes, the tRNA genes, and the genes for small nuclear RNA (snRNA) molecules.
    In the cell the 18S, 5.8S, 28S, and 5S rRNAs are structural components of ribosomes. The mRNAs am translated to produce proteins, the tRNAs bring amino acids to the ribosome to donate to the growing polypeptide chain during protein synthesis, and at least some of the snRNAs are involved in RNA processing events.

12.9 What is the Pribnow box? The Goldberg-Hogness box (TATA element)?

Answer: The Pribnow box is an important part of the prokaryotic promoter. It is located at -10 relative to the starting point of transcription and has the consensus sequence: 5'- TATAAT-3'.
    The Goldberg-Hogness box is a eukaryotic counterpart of the Pribnow box, at least for many protein-coding genes. In higher eukaryotes it is located at -25 to -30 and has the consensus sequence TATAAA. It is also referred to as the TATA element. In yeast, the TATA elements range from 40 to 120 bp upstream of the transcription starting point in contrast to the situation in higher eukaryotes.

12.10 What is an enhancer element?

Answer: An enhancer element is defined as a DNA sequence that somehow, without regard to its position relative to the gene or its orientation in the DNA, increases the amount of RNA synthesized from the gene it controls (discussed in the chapter on pp. 373-374).

12.11 A piece of mouse DNA was sequenced as follows (a space is inserted after every 10th base for ease in counting; "........." means a lot of unspecified bases.):
      AGAGGGCGGT   CCGTATCGGC   CAATCTGCTC   ACAGGGCGGA
      TTCACACGTT   GTTATATAAA   TGACTGGGCG   TACCCCAGGG
      TTCGAGTATT   CTATCGTATG   GTGCACCTGA   CT(.......)
      GCTCACAAGT   ACCACTAAGC........
What can you see in this sequence to indicate it might be all or part of a transcription unit?

Answer: The GGGCGG sequence (GC element) occurs (starts at bases 4, 34, and 76). The "CAAT box" occurs (GGCCAATCT) at base 18. A "TATA box" is seen starting at base 55. [As we shall see in later chapters, the "start translating" (ATG at base 98) and "stop translating" (TAA at base 5th from end of given sequence) signals are also here.]

CHAPTER 13

RNA MOLECULES AND RNA PROCESSING

I. CHAPTER OUTLINE

Overview of Translation
The Structure and Function of Messenger RNA
    Production of Prokaryotic mRNA
    Production of Mature Messenger RNA in Eukaryotes
The Structure and Function of Transfer RNA
    Molecular Structure of tRNA
    Transfer RNA Genes
    Biosynthesis of Transfer RNAs
Ribosomal RNA
    Structure of the Prokaryotic Ribosome in *E. coli*
    The rRNA Genes of *E. coli*
    The Ribosomal Protein Genes of *E. coli*.
    Biosynthesis of the *E. coli* Ribosome
    Structure of the Eukaryotic Ribosome
    Ribosomal RNA Genes in Eukaryotes
    Biosynthesis of Eukaryotic Ribosomes

II. IMPORTANT TERMS AND CONCEPTS

precursor RNA	spliceosomes
ribosome	snRNA *vs.* snRNP
codon	cloverleaf model
leader, coding, trailer sequence	anticodon
RNA processing	gene redundancy
colinearity	unique sequence
posttranscriptional modification	RNA ligase
intron *vs.* exon	self assembly
poly(A) tail	rDNA repeat unit
poly(A) addition site	ETS
density gradient centrifugation	ITS
Svedverg (S) units	NTS
R-loop	group I intron
hnRNA	self-splicing
splice junction	ribozyme
branch point sequence	

III. THINKING ANALYTICALLY

Once again, organization is the key to this chapter. The varieties of RNA must be distinguished with respect to their, synthesis and post-transcriptional processing, their higher order structures and associations, and their functions. Each of these categories of information must be doubled because prokaryotes and eukaryotes differ to some degree in every case. Some of the practice problems that follow should help you organize the information by RNA type and process.

IV. QUESTIONS FOR PRACTICE

A. Multiple Choice Questions

1. Which of the following contains introns? More than one answer may be correct.
    a. pre mRNA (eukaryotic)
    b. mature eukaryotic mRNA
    c. hnRNA
    d. pre tRNA
    e. pre rRNA
    f. pre mRNA (bacterial)
    g. mature bacterial mRNA

2. Which of the following has a leader or cap and/or a tail or trailer? More than one answer may be correct.
    a. prokaryotic mRNA
    b. eukaryotic mRNA
    c. pre tRNA
    d. mature tRNA
    e. rRNA

3. Which of the following mature RNAs are different from their immature precursor. More than one answer may be correct.
    a. prokaryotic mRNA
    b. eukaryotic mRNA
    c. prokaryotic tRNA
    d. eukaryotic tRNA
    e. prokaryotic rRNA
    f. eukaryotic rRNA

4. Posttranscriptional processing of all varieties of eukaryotic RNA occurs
    a. in the nucleus.
    b. in the cytoplasm.
    c. in the nucleus for some RNA and in the cytoplasm for other kinds.
    d. before the RNA transcript separates from its complementary site on DNA.

5. In which class are the RNAs most alike and also most likely to have been transcribed from repetitive DNA?
    a. mRNA
    b. tRNA
    c. rRNA

Answers: 1a,c,d,e; 2b,c; 3b,c,d,e,f; 4a; 5b

B.   Thought Questions

1. Explain why you would or would not find group I intron self splicing genes in bacteria? (Hint: Think about bacterial introns.)

2. In a typical R-looping experiment, what conclusions would you draw if you saw 4 R-loops? What about 1 R-loop? What about no R-loops? (See text pp. 383, 388.)

3. Contrast sucrose density centrifugation and CsCl density centrifugation. What is determined by by each of these techniques and how is each done? (see text Box 13.1, p. 387.)

4. Speculate on the evolution of longer introns, and the greater redundancy of tDNA in higher eukaryotes.

5. What is meant by a degenerate code? What is an ambiguous code? How do both of these concepts apply to the relationship of the mRNA codon to the tRNA anticodon? (See text pp. 395-396.)

V. ANSWERS AND SOLUTIONS TO TEXT QUESTIONS

13.1   Compare and contrast the structures of prokaryotic and eukaryotic mRNAs.

Answer: Eukaryotic mRNAs have a 5' cap and a 3' poly(A) tail. In addition most eukaryotic pre-mRNAS contain internal non-coding regions called introns that are removed as the primary transcript is processed to yield mature mRNA. Prokaryotic mRNAs do not contain introns, and they do not have the cap and tail. As a result, prokaryotic mRNAs are not substantively changed between the time they are made and the time they function in translation. Prokaryotic mRNA also differs structurally from eukaryotic mRNA in that prokaryotic mRNA is polycistronic, may be organized into operons, and undertakes the job of translation and transcription simultaneously (coupled action). Eukaryotic mRNA, on the other hand, is generally not polycistronic, is not believed to be organized into operons, and separates the activities of transcription and translation in both time and space.

13.2 Compare the structures of the three classes of RNA found in the cell.

Answer: The three classes of RNA are mRNA (text pp. 383-385), tRNA (text pp. 394-395) and rRNA (text pp. 399-403).

13.3   Many eukaryotic mRNAs, but not prokaryotic mRNAs, contain intervening sequences. What is the evidence for the presence of intervening sequences in genes? Describe how these sequences are removed during the production of mature mRNA.

Answer: See text, pp. 385-389. Intervening sequences (ivs's) are detected by R-looping experiments involving mature mRNA and DNA and/or by comparing sizes and sequences of the genes and the mature mRNAs. The ivs's are removed from precursor-mRNA by specific nuclease action. This action presumably involves a looping action, bringing together the two junction sequences and then specific cleavage and religation of the parts of the mRNA at each loop.

13.4   Discuss the posttranscriptional modifications that take place on the primary transcripts of tRNA, rRNA, and protein-coding genes.

Answer: For mRNA: 5' capping and 3' polyadenylation in eukaryotes. For tRNA: modification of a number of the bases, removal of 5' leader and 3' trailer sequences (if present), and addition of a three-nucleotide sequence (5'-CCA-3') at the 3' end of the tRNA. For rRNA in prokaryotes a pre-rRNA molecule is processed to the mature 16S, 23S, and 5S rPNAs. In eukaryotes a precursor molecule is processed to the 18S, 5.8S, and 28S rRNA molecules; 5S rRNA is made elsewhere. The large rRNAs in both types of organisms are methylated. Additionally, the same rRNAS in eukaryotes are pseudouridylated.

13.5   Distinguish between leader sequence, trailer sequence, coding sequence, intron, spacer sequence, nontranscribed spacer sequence, external transcribed spacer sequence, and internal transcribed sequence. Give examples of actual molecules in your answer.

Answer::   See text, pp. 383, 385, 386, 399- 403.

13.6 Describe the organization of the ribosomal DNA repeating unit of a higher-eukaryotic cell.

Answer: The genes for 18S, 5.8S, and 28S rRNA are arranged linearly with external transcribed spacers (ETS) flanking the 18S and 28S rDNA genes. The 5.8S rDNA, which is the middle of the three genes, is flanked on either side by internal transcribed spacers (ITS). The gene for 5S rRNA is generally found in multiple copies at different locations within the genome.

        The linear segment containing three of the four rDNA genes exists in several to many copies, in tandem repeating units separated by non-transcribed spacers (NTS). These repeating units exist in clusters within the genome, around which form the nucleoli. The number of nucleoli, and hence the number of clusters, and also the number of tandem repeats per cluster, varies widely among the eukaryotes, although all members of a species have the same number of RDNA genes and nucleoli per somatic cell nucleus.

13.7 In what way(s) is the principle of colinearity between gene and transcript violated in eukaryotes? In what sense is this principle maintained?

Answer: The eukaryotic mature transcript contains 5' and 3' sequences which are not present in the gene. In addition, the mature transcript lacks sequences which are present in the gene (introns). The principle of colinearity is nevertheless maintained in that those base sequences which are shared by the gene and the mature transcript occur in the same linear order in both.

13.8 Which of the following kinds of mutations would be likely to be recessive lethals in humans? Explain your reasoning.
    a. deletion of the U1 genes
    b. deletion within intron 2 of G-globin
    c. point mutation within the RNA polymerase III leading to loss of proofreading ability
    d. deletion of 4 bases at the end of intron 2 and 3 bases at the beginning of exon 3 in β-globin.

Answer: a. This would prevent splicing out of any sequences corresponding to introns, and would result in abnormal sequences of most proteins. This should be lethal.
    b. If splicing is not affected, this should have no phenotypic consequence.
    c. This should raise mutation rate, but not necessarily be lethal.
    d. This would prevent the splicing out of intron 2 material. Since the 5' cut could still be made, it is likely this mutation would lead to the absence of functional mRNA and the absence of β-globin. [This should produce a phenotype similar to that seen when the β-globin gene is deleted. In that case there is anemia compensated by increased production of fetal hemoglobin. The condition is called β-thalassemia.]

13.9 The diagram in the following figure shows the transcribed region of a typical eukaryotic gene:

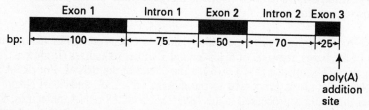

What is the size (in bases) of the fully processed, mature mRNA? Assume in your calculations a poly(A) tail of 200 As.

Answer: 376 bases = 1 (mG cap) + 100 (exon 1) = 50 (exon 3) + m200 poly (A) tail.

13.10 Which of the following could occur in a single mutational event in a human? Explain.
    a. Deletion of 10 copies of the 5S ribosomal RNA genes only.
    b. Deletion of 10 copies of the 18S rRNA genes only.
    c. Simultaneous deletion of 10 copies of the 18S, 5.8S, and 28S rRNA genes only.
    d. Simultaneous deletion of 10 copies each of the 18S, 5.8S, 28S, and 5S rRNA genes.

Answer: a. This would be possible because the 5S genes occur in clusters apart from the other rRNA genes

b. This would not be possible, because the 18S gene copies are interspersed among 5.8S and 28S genes

c. This would be possible. These three occur in tandem arrays

d. This would not be possible, because the 5S genes are located separately within the chromosomes.

13.11 During DNA replication in a mammalian cell a mistake occurs: 10 wrong nucleotides are inserted into a 28S rRNA gene, and this mistake is not corrected. What will likely be the effect on the cell?

Answer: Nothing. The cell will contain about 1,000 other copies of this gene, presumably all normal.

CHAPTER 14

THE GENETIC CODE AND THE TRANSLATION OF THE GENETIC MESSAGE

I.   CHAPTER OUTLINE

The Nature of the Genetic Code
 Deciphering the Genetic Code
 Nature and Characteristics of the Genetic Code
Translation of the Genetic Message
 Aminoacyl-tRNA Molecules
 Initiation of Translation
 Elongation of the Polypeptide Chain
 Termination of Translation
Protein Transport and Compartmentalization
 Proteins Distributed by the Endoplasmic Reticulum,
 Proteins Transported into Mitochondria and Chloroplasts
 Proteins Transported into the Nucleus

II.   IMPORTANT TERMS AND CONCEPTS

codon
cell-free protein synthesizing system
random copolymer
reading frame
degeneracy
sense codon *vs.* nonsense/
 stop/ termination codon
charging *vs.* charged tRNA
aminoacyl-tRNA synthetase
 vs. aminoacyl-tRNA
ribosome binding site
 vs Shine-Dalgarno sequence
initiation factors/complex
cap binding factor
A site *vs.* P site

elongation factors
translocation
GTP
polysome
termination factors
release factors
signal hypothesis
signal sequence
signal recognition particle/protein
signal peptidase
docking protein
cotranslational *vs.*
 posttranslational
transport
transit sequences

III.   THINKING ANALYTICALLY

At an introductory level, much of this material is descriptive and detailed. Initiation, elongation, and termination are all complex events, involving numerous enzymes and other factors that differ in prokaryotes and eukaryotes. However, in addition to spatial organization of the varieties of.RNA and factors involved in translation, it is also helpful to construct a temporal frame work to trace the order of events during construction of such items as initiation complexes and elongation sequences.

The logic involved in answering most of the questions and problems that have been posed is primarily the garden variety brand of deduction most students use all the time to solve many sorts of problems in genetics and other areas of life. However, the material concerning the history of deciphering the meaning of the codons involves probabilistic thinking. The product rule, which states that the frequency of two independent events occurring together is the product of their separate probabilities, allows one to deduce the appropriate codons that specify particular amino acids. An example follows:

Suppose that you wish to construct a synthetic mRNA by randomly combining adenine and guanine-containing ribonucleotides. What codons would be formed and what would be their relative frequency if you began with a limitless mixture of 3 adenine to 1 guanine-containing nucleotides?

Solution: First you must list all of the three-nucleotide sequences that could be made with two kinds of nucleotides and then multiply the frequency of each kind in order to determine the frequency of the combination. Remember that the chance of an adenine appearing in the synthetic mRNA is 3/4 while the chance of a guanine appearing is 1/4.

$$AAA = (3/4)^3 = 0.42$$
$$AAG = (3/4)^2(1/4) = 0.14$$
$$AGA = (3/4)^2(1/4) = 0.14$$
$$GAA = (3/4)^2(1/4) = 0.14'$$
$$AGG = (3/4)(1/4)^2 = 0.06$$
$$GAG = (3/4)(1/4)^2 = 0.06$$
$$GGA = (3/4)(1/4)^2 = 0.06$$
$$GGG = (1/4)^3 = 0.02$$

If the synthetic mRNA is translated in a cell free system containing all of the amino acids, several amino acids will be incorporated into a polypeptide. While it is not possible by this method and by one experiment to differentiate between the amino acids coded by codons of equal frequency, it should be possible to assign AAA to lysine and GGG to glycine, since those amino acids should appear in a unique ratio. Further experiments using different copolymeric ratios should help to clarify some of the ambiguous amino acid-codon assignments.

Test your use of this logic with chapter-end problems. one final note - do not forget that transcription and translation both occur in the 5'-to-3' direction. Be sure to keep this in mind when deducing base order in DNA from mRNA codon base order or tRNA anticodon base order.

IV. QUESTIONS FOR PRACTICE

A. Multiple Choice Questions

1. The eukaryotic equivalent of the Shine-Dalgarno sequence is the
   a. large ribosomal subunit.
   b. small ribosomal subunit.
   c. 5' cap of mRNA.
   d. poly(A) tail.

2. GTP is utilized during which phases of translation?
   a. initiation
   b. elongation
   c. termination
   d. initiation and elongation, but not termination

3. Which of the following are the same in both pro- and eukaryotes?
   a. elongation factors
   b. stop codons
   c. use of fMet-tRNA
   d. more than one of the above

141

4.   In *E. coli*, the codon AUG functions as the initiator codon
     a.   only if it is 8-12 bases upstream from the Shine-Dalgarno sequence.
     b.   only if it is 8-12 bases downstream from the Shine-Dalgarno sequence.
     c.   only if it is near the 5' cap.
     d.   only if the ribosome has been fully assembled.

5.   In *E. coli*, a mutation of the gene for RF1 would
     a.   have more serious consequences than a mutation in the gene for RF3.
     b.   prevent proper initiation of translation.
     c.   cause improper recognition of the Shine-Dalgarno sequence.
     d.   probably not have much effect, if all of the other factors were correct.

Answers: 1c, 2d, 3b, 4b, 5a

B.   Thought Questions

1.   The following sequence of bases in DNA contains part of a gene. Mark the first transcribed codon.
     5'-C C G C A T T A T C C G G C G G G A C C T A C T-3'
(Hint: Find the initiation sequence; watch direction!)
2.   Why do you think that the code for methionine is not degenerate? (Hint: What special role does methionine play in translation?)
3.   List the components of a cell-free protein synthesizing system and state the function of each. (See text, p. 411.)
4.   Describe three methods by which the code was deciphered, and list the disadvantages of each method. (See text, p. 412.)
5.   Why do we think that the genetic code was "settled on" very early in the evolution of cells? (Hint: Think about code universality.)

V.   ANSWERS AND SOLUTIONS TO TEXT QUESTIONS

14.1   The form of genetic information used directly in protein synthesis is (choose the correct answer):
       a.   DNA
       b.   mRNA
       c.   rRNA
       d.   ribosomes

Answer: b. mRNA

14.2   The process in which ribosomes engage is (choose the correct answer):
       a.   replication
       b.   transcription
       c.   translation
       d.   disjunction
       e.   cell division

Answer: c. translation

14.3   What are the characteristics of the genetic code?

Answer: Three-letter code, universal, nonoverlapping, degenerate, comma-free, and specific start and stop signals.

14.4   Base-pairing wobble occurs in the interaction between the anticodon of the tRNAs and the codons. On the theoretical level, determine the minimum number of tRNAs needed to read the 61 sense codons.

Answer: Because of the wobble effect, the base at the 5' end of the anticodon is less important than in the other two places. Indeed, only two different bases in that position are required because an anticodon with G at the 5' end can pair with either U or C in mRNA and an anticodon with U at the 5' end can pair with either A or G in mRNA. With meaning residing in the first two positions only, one would require 4 x 4 different tRNA molecules. With meaning only dependent on either a purine or a pyrimidine in the third position, then 4 x 4 x 2 tRNA molecules would be needed. However, UUA and UUG are stop codons. Thus, only 31 different anticodons would be required to read the sense codons. (Two more, however, would be needed to code unambiguously for Trp and Met.) The logic of this argument can be checked by consulting Figure 14.3, the chart of the genetic code. One anticodon (AAG) could read both UUU and UUG. A second (AAU) could read both UUA and UUG, etc. Each half box within the figure could be read by one anticodon. Since there are 15 half boxes of sense codons in the chart, 31 anticodons could do the reading job.

14.5   Antibiotics have been very useful in elucidating the steps of protein synthesis. If you have an artificial messenger of the sequence AUGUUUUUUUUUUUUU, it will produce the following polypeptide in a cell-free, protein-synthesizing system: fMet-Phe-Phe-Phe.... In your search for new antibiotics you find one called putyermycin, which blocks protein synthesis. When you try it with your artificial mRNA in a cell-free system, the product is fMet-Phe. What step in protein synthesis does putyermycin affect? Why?

Answer: Initiation and the formation of the first peptide bond occur, but translocation of the ribosome to the next codon is inhibited. The evidence for this result is that a dipeptide is produced. The dipeptide rules out blocks in initiation in the first step of elongation-that is, the binding of a charged tRNA molecule in the A site-and in the formation of the first peptide bond catalyzed by peptidyl transferase.

14.6   Describe the reactions involved in the aminoacylation (charging) of a tRNA molecule.

Answer: See text pp. 415, 416 and Figure 14.6.

14.7   Compare and contrast the following in prokaryotes and eukaryotes: (a) protein synthesis initiation; (b) protein synthesis elongation; (c) protein synthesis termination.

Answer: See text pp. 415-422.

14.8   Discuss the two species of methionine tRNA, and describe how they differ in structure and function. In your answer, include a discussion of how each of these tRNAs binds to the ribosome.

Answer: See text, pp. 416-419. One, fMet-tRNA, is used for initiation, and the other is used in all elongation steps. In prokaryotes, but not in eukaryotes, the methionine on the initiator tRNA is formulated on the amino group. The anticodon loop also differs between the two tRNAs, so the initiator tRNA shows 5' wobble in contrast to 3' wobble.

14.9   Random copolymers were used in some of the experiments that revealed the characteristics of the genetic code. For each of the following ribonucleotide mixtures, give the expected codons and their frequencies, and give the expected proportions of the amino acids that would be found in a polypeptide directed by the copolymer in a cell-free, protein-synthesizing system:
   a.   4 A:6 C
   b.   4 G:1 C
   c.   1 A:3 U:1 C
   d.   1 A: 1 U: 1 G: 1 C

Answer: a. 4 A:6 C

$$AAA = (\frac{4}{10})(\frac{4}{10})(\frac{4}{10}) = 0.064, \text{ or } 6.4\% \text{ Lys}$$

$$AAC = (\tfrac{4}{10})(\tfrac{4}{10})(\tfrac{6}{10}) = 0.096 \text{ or } 9.6\% \text{ Asn}$$

$$ACA = (\tfrac{4}{10})(\tfrac{6}{10})(\tfrac{4}{10}) = 0.096, \text{ or } 9.6\% \text{ Thr}$$

$$CAA = (\tfrac{6}{10})(\tfrac{4}{10})(\tfrac{4}{10}) = 0.096, \text{ or } 9.6\% \text{ Gln}$$

$$CCC = (\tfrac{6}{10})(\tfrac{6}{10})(\tfrac{6}{10}) = 0.216, \text{ or } 21.6\% \text{ Pro}$$

$$CCA = (\tfrac{6}{10})(\tfrac{6}{10})(\tfrac{4}{10}) = 0.144, \text{ or } 14.4\% \text{ Pro}$$

$$CAC = (\tfrac{6}{10})(\tfrac{4}{10})(\tfrac{6}{10}) = 0.144, \text{ or } 14.4\% \text{ His}$$

$$ACC = (\tfrac{4}{10})(\tfrac{6}{10})(\tfrac{6}{10}) = 0.144, \text{ or } 14.4\% \text{ Thr}$$

In sum,6.4% Lys, 9.6% Asn, 9.6% Gln, 36.0% Pro, 24.0% Thr, and 14.4% His.

b. 4G:1C

$$GGG = (\tfrac{4}{5})(\tfrac{4}{5})(\tfrac{4}{5}) = \qquad 0.512, \text{ or } 51.2\% \text{ Gly}$$

$$GGC = (\tfrac{4}{5})(\tfrac{4}{5})(\tfrac{1}{5}) = \qquad 0.128, \text{ or } 12.8\% \text{ Gly}$$

$$CCG = (\tfrac{4}{5})(\tfrac{1}{5})(\tfrac{4}{5}) = \qquad 0.128, \text{ or } 12.8\% \text{ Ala}$$

$$CGG = (\tfrac{1}{5})(\tfrac{4}{5})(\tfrac{4}{5}) = \qquad 0.128, \text{ or } 12.8\% \text{ Arg}$$

$$CCC = (\tfrac{1}{5})(\tfrac{1}{5})(\tfrac{1}{5}) = \qquad 0.008, \text{ or } 0.8\% \text{ Pro}$$

$$CCG = (\tfrac{1}{5})(\tfrac{1}{5})(\tfrac{4}{5}) = \qquad 0.032, \text{ or } 3.2\% \text{ Pro}$$

$$CGC = (\tfrac{1}{5})(\tfrac{4}{5})(\tfrac{1}{5}) = \qquad 0.032, \text{ or } 3.2\% \text{ Arg}$$

$$GCC = (\tfrac{4}{5})(\tfrac{1}{5})(\tfrac{1}{5}) = \qquad 0.032, \text{ or } 3.2\% \text{ Ala}$$

In sum, 64.0% Gly, 16.0% Ala, 16.0% Arg, and 4.0% Pro.

c. 1 A:3 U:1 C; the same logic is followed here, using 1/5 as the fraction for A, 3/5 for U,and 1/5 for C.

AAA = 0.008, or 0.8% Lys
AAU = 0.024, or 2.4% Asn
AUA = 0.024, or 2.4% Ile
UAA = 0.024, or 2.4% chain terminating
AUU = 0.072, or 7.2% Ile
UAU = 0.072, or 7.2% Tyr
UUA = 0.072, or 7.2% Leu
UUU = 0.216, or 21.6% Phe
AAC = 0.008, or 0.8% Asn
ACA = 0.008, or 0.8% Thr
CAA = 0.008, or 0.8% Gln
ACC = 0.008, or 0.8% Thr
CAC = 0.008, or 0.8% His
CCA = 0.008, or 0.8% Pro
CCC = 0.008, or 0.8% Pro
UUC = 0.072, or 7.2% Phe

UCU = 0.072, or   7.2% Ser
CUU = 0.072, or   7.2% Leu
UCC = 0.024, or   2.4% Ser
CUC = 0.024, or   2.4% Leu
CCU = 0.024, or   2.4% Pro
UCA = 0.024, or   2.4% Ser
UAC = 0.024, or   2.4% Tyr
CUA = 0.024, or   2.4% Leu
CAU = 0.024, or   2.4% His
AUC = 0.024, or   2.4% Ile
ACU = 0.024, or   2.4% Thr

In sum, 0.8% Lys, 3.2% Asn, 12.0% Ile, 2.4% chain terminating, 9.6% Tyr, 19.2% Leu, 28.8% Phe, 4.0% Thr, 0.8% Gln, 3.2% His, 4.0% Pro, and 12.0% Ser. The likelihood is that the chain would not belong because of the chance of the chain-terminating codon.

d. 1 A:1 U:1 G:1 C; all 64 codons will be generated. The probability of each codon is 1/64, so there is a 3/64 chance of the codon being a chain-terminating codon. With those exceptions the relative proportion of amino acid incorporation is directly dependent on the codon degeneracy for each amino acid, and that can be determined by inspecting the code word dictionary.

14.10   Other features of the reading of mRNA into proteins being the same as they are now (i.e., codons must exist for 20 different amino acids), what would be the minimum word (codon) size if the number of different bases in the mRNA were, instead of four:
  a. two
  b. three
  c. five

Answer: *Word size*   *Number of combinations*

      a. 5 $2^5 = 32$
      b. 3 $3^3 = 27$
      c. 2 $5^2 = 25$

(The minimum word size must uniquely designate twenty amino acids.)

14.11   Suppose that at stage A in the evolution of the genetic code only the first two nucleotides in coding triples led to unique differences and that any nucleotide could occupy the third position. Then, suppose there was a stage B in which differences in meaning arose depending upon whether a purine (A or G) or pyrimidine (C or T) was present at the third position. Without reference to the number of amino acids or multiplicity of tRNA molecules, how many triplets of different meaning can be constructed out of the code at stage A? at stage B?

Answer: Stage A: $4^2 = 16$
      Stage B: $4^2 \times 2 = 32$

4.12   A gene makes a polypeptide 30 amino acids long containing an alternating sequence of phenylalanine and tyrosine. What are the sequences of nucleotides corresponding to this sequence in.
  a. the DNA strand which is read to produce the mRNA, assuming Phe = UUU and Tyr = UAU in mRNA.
  b. the DNA strand which is not read.
  c. tRNA.

Answer: a. AAA ATA AAA ATA etc.
      b. TTT TAT TTT TAT etc.
      c. AAA for Phe and AUA for Tyr

14.13 A segment of a polypeptide chain is Arg-Gly-Ser-Phe-Val-Asp-Arg. It is encoded by the following segment of DNA:

```
————GGCTAGCTGCTTCCTTGGGGA————
    ||||||||||||||||||||||
————CCGATCGACGAAGGAACCCCT————
```

Which strand is the template strand? Label each strand with its correct polarity (5' and 3').

Answer: The template strand is the one that is read to produce the mRNA. Since mRNA is made in the 5'-to-3' direction, the template DNA strand is the 3'-to-5' complement of the mRNA. Arg is specified by codons CGU, CGC, CGA, CGG, AGA, and AGG (see Fig. 14.3). The AGG codon is found on the top DNA strand reading from right to left at the right end. The other codons continue on that strand as would be predicted. Thus, the top DNA strand, reading from right to left has the *same* polarity as the mRNA – it is the non-template strand. The bottom DNA strand, therefore, is the template strand and the polarities area as follows:

```
3'-GGCTAGCTGCTTCCTTGGGGA-5'    Non-template strand

5'-CCGATCGACGAAGGAACCCCT-3'    Template strand
```

↓ Transcription

```
3'-GGC TAG CTG CTT CCT TGG GGA-5'
```

↓ Translation

```
Arg Asp Val Phe Ser Gly Arg
```

14.14 Two populations of RNAs are made by the random combination of nucleotides. In population A the RNAs contain only A and G nucleotides (3A:1G), while in population B the RNAs contain only A and U nucleotides (3A:1U). In what ways other than amino acid content will the proteins produced by translating the population A RNAs differ from those produced by translating the population B RNAs?

Answer: No chain terminating codons can be produced from only As and Gs. On the other hand, the stop codon UAA can be made from As and Us. Therefore the population A proteins will be longer than those from population B. Most of the population B proteins will be free in solution rather than attached to ribosomes.

14.15 In *E. coli* a particular tRNA normally has the anticodon 5'-GGG-3', but because of a mutation in the tRNA gene, the mutant tRNA has the anticodon 5'-GGA-3'.
    a. What amino acid would this tRNA carry?
    b. What codon would the normal tRNA recognize?
    c. What codon would the mutant tRNA recognize?
    d. What would be the effect of the mutation on the proteins in the cell?

Answer: a. Proline
        b. CCC
        c. UCC
        d. Proline would be inserted in places where serine belongs in every kind of protein made by the cell. Also, proline would fail to be inserted at CCC codon, halting translation at that point.

14.16 A particular protein found in *E. coli* normally has the N-terminal sequence Met-Val-Ser-Ser-Pro-Met-Gly-Ala-Ala-Met-Ser.... In a particular cell a mutation alters the anticodon of a particular

146

tRNA from 5'-GAU-3' to 5'-CAU-3'. What would be the N-terminal amino acid sequence of this protein in the mutant cell? Explain your reasoning.

Answer: Met-Val-Ser-Ser-Pro-Ile-Gly-Ala-Ala-Ile-Ser.....(In fact either of the Ile residues might be replaced by Met in a particular molecule.) The normal tRNA recognizes the AUC codon, and therefore carries Ile. The mutant tRNA will recognize the codon AUG, and insert Ile there (although the normal tRNA-Met will compete for these sites.) The N terminal Met will not be replaced by Ile since it requires the special tRNA-fMet for initiation to occur.

14.17 The gene encoding an E. *coli* tRNA containing the anticodon 5'-GUA-3' mutates so that the anticodon now is 5'-UUA-3'. What will be the effect of this mutation? Explain your reasoning.

Answer: The normal tRNA recognized the codon UAC, and so must have carried the amino acid tyrosine. The altered anticodon will recognize the codon UAA. UAA is a chain termination codon. If an tyrosine is inserted when UAA appears in the A site on the ribosome, presumably read-through will occur, and extra amino acids will be added onto the end of the protein. Some chains are terminated by UAG or UGA and will not be affected.

14.18 The normal sequence of the coding region of a particular mRNA is shown in the following figure, along with several mutant versions of the same mRNA. Indicate what protein would be formed in each case. (... = many [a multiple of 31 unspecified bases.)

normal:	AUGUUCUCUAAUUAC(...)AUGGGGUGGGUGUAG
mutant *a*:	AUGUUCUCUAAUUAG(...)AUGGGGUGGGUGUAG
mutant *b*:	AGGUUCUCUAAUUAC(...)AUGGGGUGGGUGUAG
mutant *c*:	AUGUUCUCGAAUUAC(...)AUGGGGUGGGUGUAG
mutant *d*:	AUGUUCUCUAAAUAC(...)AUGGGGUGGGUGUAG
mutant *e*:	AUGUUCUCUAAUUC(...)AUGGGGUGGGUGUAG
mutant *f*:	AUGUUCUCUAAUUAC(...)AUGGGGUGGGUGUGG

Answer: Normal: Met-Phe-Ser-Asn-Tyr-(...)-Met-Gly-Trp-Val.
Mutant *a*: Met-Phe-Ser-Asn
Mutant *b*: starts out late to give Met-Gly-Trp-Val.
Mutant *c*: Met-Phe-Ser-Asn-Tyr-(...)-Met-Gly-Trp-Val.
Mutant *d*: Met-Phe-Ser-Lys-Tyr-(...)-Met-Gly-Trp-Val.
Mutant *e*: Met-Phe-Ser-Asn-Ser-(...)-Trp-Gly-Gly-Trp-.....
Mutant *f*: Met-Phe-Ser-Asn-Tyr-(...)-Met-Gly-Trp-Val-Trp....
(No stop codons for *e* and *f*, so protein continues.)

14.19 The normal sequence of a particular protein is given below, along with several mutant versions of it. For each mutant, explain what mutation occurred in the coding sequence of the gene.

Normal:	Met-Gly-Glu-Thr-Lys-Val-Val-( )-Pro
Mutant 1:	Met-Gly
Mutant 2:	Met-Gly-Glu-Asp
Mutant 3:	Met-Gly-Arg-Leu-Lys
Mutant 4:	Met-Arg-Glu-Thr-Lys-Val-Val-( )-Pro

Answer: In mutant 1 a GC base pair changed to a TA base pair. The normal C gave rise to a G in the 3rd mRNA codon GAG or GAA (Glu). The mutant mRNA has UAA or UAG (stop).
In mutant 2 a GC base pair was inserted, so that the ACU codon for Thr became GAC, and the leftover U joined with AA from the Lys codon to form a UAA stop signal.
In mutant 3 a deletion of a single base pair occurred, leading either to the deletion of the first base (G) of the third codon, or of any base of the second codon. Evidently Glu was coded by GAG and Thr by ACU. The new 3rd codon is now AGA (Arg), followed by CUA (CU from ACU and A from the following AAA, etc).
In mutant 4 a GC pair changed to a CG pair, such that the second codon became CGG (Arg) instead of GGG (Gly).

14.20 In the recessive condition in humans known as Sickle Cell Disease (SCD), the β-globin moiety of hemoglobin is found to be abnormal. The only difference between it and the normal β-globin is that the 6th amino acid from the N-terminal is valine, whereas the normal β-globin has glutamic acid at this position. Explain how this occurred.

Answer: The codons for glutamic acid are GAA and GUG. The codons for valine are GUU, GUC, GUA, and GGG. The simplest explanation is that the 17th base pair in the coding region of the gene changed from AT to TA, so that the 6th codon of the mRNA, which is supposed to be GAA or GAG, became GUA or GUG.

14.21 Antibiotics have been useful in determining whether cellular events depend on transcription or translation. For example, actinomycin D is used to block transcription, and cycloheximide (in eukaryotes) is used to block translation. In some cases, though, surprising results are obtained after antibiotics are administered. The addition of actinomycin D, for example, may result in an increase and not a decrease in the activity of a particular enzyme. Discuss how this result might come about.

Answer: A probable reason is that actinomycin D might block the transcription of a gene that codes for an inhibitor of an enzyme activity.

CHAPTER 15

RECOMBINANT DNA TECHNOLOGY AND THE MANIPULATION OF DNA

I. CHAPTER OUTLINE

Restriction Endonucleases
Cloning Vectors and Cloning
    Plasmid Cloning Vectors
    Bacteriophage Lambda Cloning Vectors
    Cosmid Cloning Vectors
    Shuttle Vectors
Construction of Genomic Libraries, Chromosome Libraries, and cDNA Libraries
    Genomic Libraries
    Chromosome Libraries
    cDNA Libraries
Identifying Specific Cloned Sequences in cDNA Libraries and Genomic Libraries
    Identifying Specific Cloned Sequences in a cDNA Library
    Identifying Specific Cloned Sequences in a Genomic Library
    Identifying Specific DNA Sequences in Libraries Using Heterologous Probes
Identifying Genes in Libraries by Complementation of Mutations
    Identifying Genes or cDNAs in Libraries Using Oligonucleotide Probes
Techniques for the Analysis of Genes and Gene Transcripts
    Restriction Enzyme Analysis of Genes
    Restriction Enzyme Analysis of Cloned DNA Sequences
    Analysis of Gene Transcripts
DNA Sequence Analysis
    Maxam-Gilbert DNA Sequencing
    Dideoxy (Sanger) DNA Sequencing
    Analysis of DNA Sequences
Polymerase Chain Reaction (PCR)
Applications of Recombinant DNA Technology
    Analysis of Biological Processes
    Diagnosis of Human Genetic Diseases
    DNA Fingerprinting
    Human Gene Therapy
    Human Genome Project
    Commercial Products
    Genetic Engineering of Plants

II. IMPORTANT TERMS AND CONCEPTS

molecular cloning	flow cytometry
cloning vector	linker
restriction endonuclease	expression vector
restriction site	autoradiography

isoschizomers

blunt end *vs.* sticky end

plasmid

dominant selectable marker

polylinker/multiple cloning site

lambda replacement vector

disposable segment

ori site *vs. cos* site

genomic library

cDNA

open reading frame

partial digestion

Ti plasmid/T-DNA

nick translation

random primer method

plaque lift

heterologous probe

gel electrophoresis

Southern blotting

restriction map

Northern blotting

Dideoxy sequencing

Maxam-Gilbert sequencing

polymerase chain reaction

human genome project

III. THINKING ANALYTICALLY

The material in this chapter is illustrative of the state of modern genetics. It presents many of the areas and issues that are on the cutting edge of progress in the discipline. In addition, this material illustrates the key role played by technology in new discoveries and developments and it also shows how basic theoretical science can be put to applied uses in the diagnosis and treatment of human diseases and in the search for solutions to other human problems. Thoroughly learning this material requires a clear understanding of the logic and process of many different techniques, a daunting task-. that would be assisted by actually doing, or at least observing, some of the procedures that are discussed.

This chapter also presents the student with material at a much more sophisticated level of abstraction than is required for much of biology - one that approaches the level seen in the physical sciences of chemistry and physics. In its simplest form, the key question is how one manipulates things that one can not sense directly with eyes, ears, nose, touch, etc. Each of the techniques explained in this chapter is based on a "model" of reality, on the belief that we have of the correct structure, function, and activity of DNA, RNA, cells, and viruses under certain conditions of temperature, growth media, and other environmental parameters. Each time one performs one of the techniques outlined in Chapter 15 and gets results that appear to be reasonable, one is really testing the validity of models that geneticists have arrived at for the structure and action of various molecules and cellular processes. For example, we can not actually see the base sequence in a length of DNA, but we have constructed a model of its structure based on prior experimental evidence. If that model is correct, then a technique such as the Sanger sequencing method should work. As you consider each of the methodologies presented in Chapter 15, ask yourself which basic models are being tested. Pretend that the model is wrong in some respect, and ask yourself how the results of the technique might be altered.

IV. QUESTIONS FOR PRACTICE

A. Multiple Choice Questions

For questions 1-5 match each of the techniques listed on the right with the statement that describes how or for what purpose it is used.

PURPOSE

1. To isolate a particular cDNA clone in a cDNA library.
2. To separate mRNA from a mixture of m, r, and tRNA prior to construction of cDNA.
3. To determine whether a particular mRNA is present in a cell at a specific stage of development.
4. To determine gene homologies of intron organization.
5. To produce relatively large DNA fragments appropriate for a genomic DNA library.

TECHNIQUES

a. restriction mapping
b. partial digestion with restriction enzyme that recognizes 4 base pair fragments.
c. base pairing poly(dT) to poly(A) tails
d. Northern blotting.
e. Use of expression vectors to produce a particular protein.

6. Antibiotic resistance markers are important in the use of plasmid cloning vectors
   a. because the plasmid must show resistance in order to accept inserted DNA.
   b. so that any previously sensitive bacterium they have taken up a such a plasmid can be recognized.
   c. so that one can be sure that both the ori site and the *cos* site are present.
   d. so that the resistance gene.can be cut by a restriction enzyme.

Answers: 1e, 2c, 3d, 4a, 5b, 6b

B. Thought Questions

1. Contrast the *cos* site with the ori site. (See text pp. 439, 441.)
2. Contrast the nature and use of cosmids with that of plasmids. (See text pp. 439-441)
3. What kinds of organisms naturally make restriction enzymes? Of what use are they to the organisms that naturally make them? (See text p 432)
4. When would you use a restriction enzyme that recognized an 8 base pair, rather than a 4 base pair, sequence? (See text p. 434.)
5. Construct a restriction map of a length of DNA using the following data (see text, pp. 453-456.):

uncut DNA	10 kb
DNA cut with *Eco* RI	2 kb, 5 kb
DNA cut with *Bam* HI	5 kb, 8 kb
DNA cut with *Eco* RI+ *Bam* HI	2 kb, 3 kb, 5,kb

6. What is the sequence of bases in a length of DNA that gave the following electrophoretic results as a result of Sanger dideoxy sequencing? Is this piece of DNA the template strand or its complement? (See text pp. 460-462.)

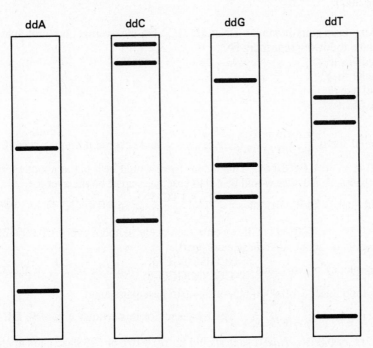

7. If you were given the power to spend the money that the human genome project will cost, would you do so?

V. ANSWERS AND SOLUTIONS TO TEXT QUESTIONS

15.1 A new restriction endonuclease is isolated from a bacterium. This enzyme cuts DNA into fragments that are, on the average, 4096 base pairs long. Like all other known restriction enzymes, the new one recognizes a sequence in DNA that has twofold rotational symmetry. From the information given, how many base pairs of DNA constitutes the recognition sequence for the new enzyme?

Answer: At any one position in the DNA, there are four possibilities for the base pair A-T, T-A,G-C, and C-G. Therefore the length of the base-pair sequence that the enzyme recognizes is given by the power to which four must be raised to equal (or approximately equal) the average size of the DNA fragment produced by enzyme digestion. The answer in this case is 6; that is, 4 to the power of $6 = 4096$. If the enzyme instead recognized a four base-pair sequence, the average size of the DNA fragment would be 4 to the power of $4 = 256$.

15.2 An endonuclease called *Avr*II ("a-v-r-two") cuts DNA whenever it finds the sequence 5'-CCTAGG-3' 3'-GGATCC-5'. About how many cuts would *Avr*II make in the human genome, which is about 3 x 10^9 base pairs tong and about 40% GC?

Answer: The sequence 5'-CCTAGG-3' 3'-GGATCC-5' should occur in $(0.2)(0.2)(0.3)(0.3)(0.2)(0.2) = 0.000144$ of all groups of six. (This calculation assumes a random distribution of bases, and so is only approximately correct for any natural DNA.) There are about 3 x 10^9 groups of six in the human genome, so the total number of *Avr*II cuts would be expected to be $(0.000144)(3$ x $10^9)$, or 432,000.

15.3 About 40% of the base pairs in human DNA are GC. On the average, how far apart (in terms of base pairs) will the following sequences be?
  a. Two *Bam*HI sites?
  b. Two *Eco*RI sites?
  c. Two *Not*I sites?
  d. Two *Hae*III sites?

Answer: a. A *Bam*HI site is 5'-GGATCC-3' 3'-CCTAGG-5' these sites would occur at a frequency of $(0.2)^4(0.3)^2$, or 0.000144 in 40% GC DNA. This means there would be a cut at about every $1/0.000144 = 694$ bases, so the sites would be 6,940 base pairs apart on the average.

b. The frequency of *Eco*RI sites 5'-GAATTC-3' 3'-CTTAAG-5' in 40% GC DNA would be about $(0.2)^4(0.3)^4$, or 0.000324. Sites would then occur at about every $1/0.000324$ base, so the sites would be about 3,086 base pairs apart.

c. The frequency of *Not*I sites 5'-GCGGCCGC-3' 3'-CGCCGGCG-5' would be $(0.2)^8$, or 0.0000026. These sites would thus be 1/0.0000025, or 384,615 base pairs apart.

d. *Hae*III recognizes 5'-GGCC-3' 3'-CCGG-5'. The frequency of this sequence in 40% GC DNA should be $(0.2)^4$, or 0.0016. *Hae*III sites should be 1/0.0016, or 625 base pairs apart.

15.4 What are the features of plasmid cloning vectors that make them useful for constructing and cloning recombinant DNA molecules?

152

Answer: Plasmid cloning vectors contain three essential features:
1. An origin of replication to permit replication in the host organism;
2. A dominant selectable marker to permit host cells containing the plasmid to be selected from host cells lacking the plasmid. Bacterial vectors typically have antibiotic-resistance genes as selectable markers, e.g., amp^R;
3. Unique restriction sites into which foreign DNA fragments can be cloned.

15.5   Genomic libraries are important resources for isolating genes of interest and for studying the functional organization of chromosomes. List the steps you would use to make a genomic library of yeast in a lambda vector.

Answer: Genomic libraries were discussed on pp. 442-444 and cloning in a lambda vector by replacing a central section of the λ chromosome with foreign DNA (see text pp. 442, 448). The steps to make a yeast genomic library are:
1. Isolate high-molecular-weight DNA from yeast nuclei;
2. Perform a partial digest of the DNA with a restriction enzyme that cuts frequently (e.g., *Sau*3A), and isolate DNA fragments of the correct size range for cloning in the λ vector by sucrose gradient centrifugation;
3. Remove the central section of an appropriate λ vector by digestion with *Bam*HI;
4. Ligate the left and right λ arms to the yeast DNA fragments – the *Sau*3A and *Bam*HI sticky ends are complementary. (See text, p. 448).
5. Package the recombinant DNA molecules *in vitro* into λ phage particles;.
6. Infect *E. coli* cells with the λ phage population, and collect progeny phages produced by cell lysis. These phages represent the yeast genomic library.

15.6   The human genome contains about 3×10^9 bp of DNA. How many 40 kb pieces would you have to clone into a library if you wanted to be 90% certain of including a particular sequence?

Answer: The probability of having any sequence represented in a genomic library is given by $N = \dfrac{\ln(1\text{-}P)}{\ln(1\text{-}f)}$ (p.444), where N is the necessary number of DNA molecules, P is the probability desired, and f is the fractional proportion of the genome in a single recombinant DNA molecule. P in this case is 0.9, and f is 40,000/3,000,000,000. So

$$N = \frac{\ln(1\text{-}0.9)}{\ln(1\text{-}[40,000/3,000,000,000])}$$ giving 172,693 DNA pieces required.

15.7   Suppose you wanted to produce human insulin peptide hormone) by cloning. Assume that this could be done by inserting the human insulin gene into a bacterial host, where, given the appropriate conditions, the human gene would be transcribed then translated, into human insulin. Which do you think it would be best to use as your source of the gene, human genomic insulin DNA or a cDNA copy of this gene? Explain your choice.

Answer: Human genomic DNA contains introns. mRNA transcribed off needs to be processed, before it can be translated not contain these processing enzymes. So in the bacterium, even if the human mRNA were translated, the protein it would code for would not be insulin. cDNA is the complementary copy of a functional mRNA molecule. So when this is the template, its mRNA transcript will be functional and when translated, human (pro-)insulin will be synthesized.

15.8   You are given a genomic library of yeast prepared in a bacterial plasmid vector. You are also given a cloned cDNA for human actin, a protein which is conserved in protein sequences among eukaryotes. Outline how you would use these resources to attempt to identify the yeast actin gene.

Answer: Screening plasmid libraries for specific DNA sequences was described on pp. 448-450 and illustrated in Figure 15.15. The steps would be

1. Replicate the plasmid library (e.g., from microtiter-dishes) to Petri dishes with nitrocellulose filters placed on top of the growth medium;.
2. After colonies have grown on the filter, remove the filter and treat it in solutions to lyse the cells, denature the DNA to single strands, and fix the single strands to the filter;
3. Place the filter in a bag with hybridization solution and radioactively-labeled, single-stranded cDNA probe;
4. Wash the filter free of unbound radioactive probe/, dry it, and prepare an autoradiogram. Dark spots indicate where the probe bound to cloned DNA in a bacterial colony, identifying a cloned DNA fragment carrying the putative yeast actin gene.

15.9 Restriction endonucleases are used to construct restriction maps of linear or circular pieces of DNA. The DNA is usually produced in large amounts by recombinant DNA techniques. The generation of restriction maps is similar to the process of putting the pieces of jigsaw puzzle together. Suppose we have a circular piece of double-stranded DNA that is 5000 base pairs long. If this DNA is digested completely with restriction enzyme 1, four DNA fragments are generated: fragment *a* is 2000 base pairs long; fragment *b* is 1400 base pairs long; *c* is 900 base pairs long; and *d* is 700 base pairs long. If, instead, the DNA is incubated with the enzyme for a short time, the result is incomplete digestion of the DNA; not every restriction enzyme site in every DNA molecule will be cut by the enzyme, and all possible combinations of adjacent fragments can be produced. From an incomplete digestion experiment of this type, fragments of DNA were produced from the circular piece of DNA, which contained the following combinations of the above fragments: *a-d-b, d-a-c, c-b-d, a-c, d-a, d-b,* and *b-c*. Lastly, after digesting the original circular DNA to completion with restriction enzyme I, the DNA fragments were treated with restriction enzyme II under conditions conducive to complete digestion. The resulting fragments were: 1400, 1200, 900, 800, 400, and 300. Analyze all the data to locate the restriction enzyme sites as accurately as possible.

Answer:

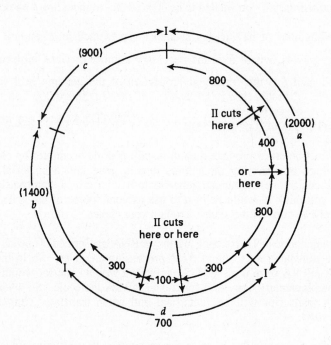

15.10 A piece of DNA 5000 bp long is digested with restriction enzymes A and B, singly and together. The DNA fragments produced we're separated by DNA electrophoresis and their sizes were calculated, with the following results:

DIGESTION WITH

A	B	A + B
2100 bp	2500 bp	1900 bp
1400 bp	1300 bp	1000 bp
1000 bp	1200 bp	800 bp
500 bp		600 bp
		500 bp
		200 bp

Each A fragment was extracted from the gel and digested with enzyme B, and each B fragment was extracted from the gel and digested with enzyme A. The sizes of the resulting DNA fragments were determined by gel electrophoresis, with the following results.

A FRAGMENT	FRAGMENTS PRODUCED BY DIGESTION WITH B	B FRAGMENT	FRAGMENT PRODUCED BY DIGESTION WITH A
2100 bp	→ 1900 , 200 bp	2500 bp	→ 1900, 600 bp
1400 bp	→ 800 , 600 bp	1300 bp	→ 800, 500 bp
1000 bp	→ 1000 bp	1200 bp	→ 1000, 200 bp
500 bp	→ 500 bp		

Construct a restriction map of the 5000 bp DNA fragment.

Answer: The restriction map is:

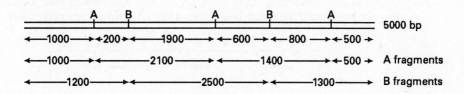

The map is built by considering the large fragments first. For example, the 250 bp B fragment is cut with A to produce 1900-bp and 600-bp fragments. The 2100-bp A fragment is cut with 8 to produce the same 1900-bp fragment, and a 200-bp fragment., Thus the 200-bp and 600-bp fragments must be on opposite sides of the 1900-bp fragment. The map is extended in a step-by-step fashion, by next considering other cuts which produced 200-bp fragments, 600-bp fragments, and so on.

15.11 Draw the banding pattern you would expect to see on a DNA-sequencing gel if you applied the Maxam and Gilbert DNA-sequencing method to the following single-stranded DNA fragment (which is labeled at the 5' end with ^{32}P): 5'^{32}P-A-A-G-T-C-T-A-C-G-T-A-T-A-G-G-C-C3'.

Answer:

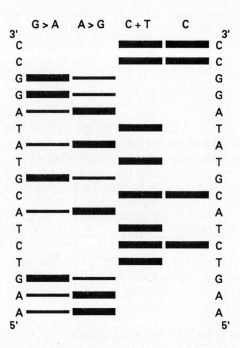

15.12 DNA was prepared from small samples of white blood cells from a large number of people. Ten different patterns were seen when these DNAs were all digested with *Eco*R1, then subjected to electrophoresis and Southern blotting. Finally, the blot was probed with a radioactively labeled cloned human sequence. The figure below shows the ten DNA patterns taken from ten people.

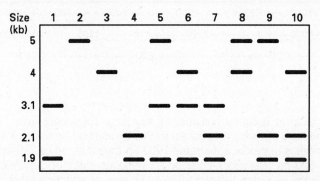

   a. Explain the hybridization patterns seen in the ten people in terms of variation in *Eco*R1 sites.
   b. If the individuals whose DNA samples are in lanes 1 and 6 on the blot were to produce offspring together, what bands would you expect to see in DNA samples from these offspring?

Answer: a. Two different variations occur in this region. A simple way to explain the patterns seen would be to postulate two variant and two invariant *Eco*RI sites. The two invariant ones are 5 kb apart. Call these two sites E1 and E2. Site E3, one of the variant sites, is 1 kb from E1 in the direction of E2. The region of homology to the probe is entirely contained between E2 and E3. Thus when the E3 site is present, the 1-kb piece representing the E1 to E3 interval is not seen on the gel. The fourth site, E4, is about midway between E3

156

and E2, dividing this 4-kb region into a 2.1 and a 1.9-kb segment. Individuals 1 through 4 are homozygotes. Individual 1 has cuts at E1, E2 and E4 in both chromosomes. #2 has cuts only at E1 and E2 in both chromosomes, #3 has cuts at E2 and E3 (and presumably at E1) in both chromosomes, and #4, has cuts at all four sites in both chromosomes. Individuals 5 through 10 are heterozygotes. #5 has one chromosome like #1 and one like #2. #6 has one like #1 and one like #3, #7 has one like #1 and one like #4. Individual #8 has one chromosome like #2 and one like #3, #9 has one like #2 and one like #4, and individual #10 has one like #3 and one like #4.

b.   If #1 produced offspring with #6, all offspring would receive the E1, E2, E4 chromosome from individual #1. Half of them would receive this same kind of chromosome from #6, and would thus be homozygous. Their DNA would look just like that of #1. The other half of the offspring would get the E1, E2, E3 chromosome from #6. They would be heterozygous, and their DNA would look just like that of #6.

15.13   Filled symbols in the pedigree below indicate people with a rare autosomal dominant genetic disease.

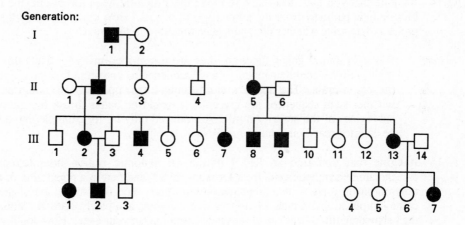

DNA samples were prepared from each of the individuals in the pedigree. The samples were digested with a restriction enzyme, electrophoresed, blotted, and probed with a cloned human sequence called DS12-88, with the results shown in the following figure. Do the data in these two figures support the hypothesis that the locus for the disease that is segregating in this family is linked to the region homologous to DS12-88? Make your answer quantitative, and explain your reasoning.

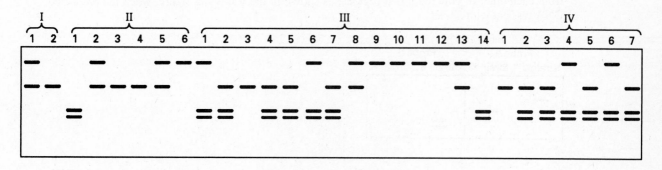

Answer: I-1 is heterozygous for the RFLP and the disease. Inspection of the blot figure shows three different "haplotypes" (DNA types) are segregating in this family. Let us name them A, B,

157

and C in decreasing order of size. Let us use d for the normal allele of the disease gene, and D for the allele causing the disease. I-1 is then Dd and AB, but we don't know whether he is DA/dB or DB/dA, so we cannot tell which of his offspring are recombinant and which are nonrecombinant. We can tell that I-2 is dB/dB, so we know the genotypes of II-2 and II-5 are both DA/dB. II-2 has six offspring. All of these except III-6 have received either DA or dB, and are non-recombinants. III-6 has received d and A, and is a recombinant. Of the six offspring of II-5, III-12 and III-13 have received recombinant chromosomes (dA for III12 and DB for III-13. The three offspring of III-2 all received non-recombinant chromosomes, as did three of the four offspring of III-13. IV-6, however, received a recombinant chromosome (dB). Overall there are 19 individuals whose recombinant *vs.* non-recombinant status can be ascertained. Of these, four are recombinant. $4/19 = 0.21$, which is less than the 0.5 expected from independent assortment. But is it *significantly* less? To find out, we must do the $\chi 2$ test. Independent assortment predicts $19/2 = 9.5$ recombinants and the same number of non-recombinants. The value of $\chi 2$ is thus 6.368, with one degree of freedom. The $\chi 2$ table (Chapter 5, p.139) indicates this difference is significant, so we can conclude the two loci are linked.

15.14   Imagine that you have been able to clone the structural gene for an enzyme in a catecholamine biosynthetic pathway from the adrenal gland of rats. How could you use this cloned DNA as a probe to determine whether this same gene functions in the brain?

Answer: When structural genes function, they are transcribed mRNA. Thus this gene functions in brain, mRNA complementary in base sequence to one strand of the cloned DNA should be present in brain. The northern blot technique (see pp. 456-458) could be used to determine whether such sequences are present. It would be better to ask the question about defined subregions of the brain. An mRNA produced only in, for example, hypothalamus might not be detected readily in whole brain RNA.

15.15   Imagine that you find an RFLP in the rat genomic region homologous to your cloned catecholamine synthetic gene from Question 15.14, and that in a population of rats displaying this polymorphism there is also a behavioral variation. You find that some of the rats are normally calm and placid, but others are hyperactive, nervous, and easily startled. Your hypothesis is that the behavioral difference seen is caused by variations in your gene. How could you use your cloned sequence to test this hypothesis?

Answer: You could study DNA from normally and oddly behaving rats, and ask whether one RFLP type is limited to a particular behavioral category. Whatever the answer, you could then do crosses between rats (making various behavioral combinations) and determine whether or not there is cosegregation between the behavioral phenotype and the RFLP. Cosegregation, if observed, is not by itself evidence for a causal relationship (it could simply indicate linkage), but the absence of cosegregation is certainly evidence against a causal relationship.

15.16   The PCR technique was used to amplify two genomic regions (homologous to probes A and B) in DNA from *Neurospora*. Two strains of opposite mating type (strains J and K) were found to differ from each other in both amplified regions, as shown in the following figure, when the amplified DNA was cut with *Eco*Rl.

Strain J was crossed to strain K, and 100 asci resulting from the cross were dissected, and the PCR reaction was done on DNA from the individual spores in each ascus. Six different patterns of distribution of DNA types within asci were seen as shown in the following figure (next page):

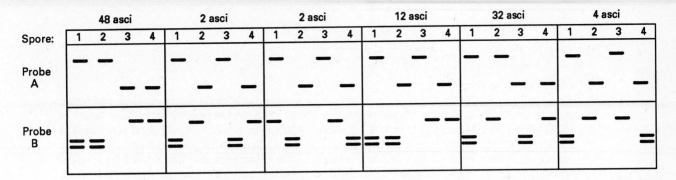

Only four patterns are shown for each ascus. Remember that spores 1 and 2 in the ascus are ordinarily identical, as are spores 3 and 4, 5 and 6, and 7 and 8. In Figure 15.E the band pattern displayed by spores 1 and 2 is indicated in the lane designated "1". The pattern shown by spores 3 and 4 is in lane 2, the pattern shown by spores 5 and 6 is in lane 3, and the pattern shown by spores 7 and 8 is in lane 4. Draw a map showing the relationships among the region detected by probe A, the region homologous to probe B, and any relevant centromeres.

Answer: Let us use the symbol "A" to represent the heavier molecular weight band detected by probe A, and "a" to represent the lighter molecular weight band. Similarly, let us use "B" to symbolize the heavier molecular weight band detected by probe B, and "b" to symbolize the doublet of lighter bands. Then the genotype of strain J can be designated as Ab, and that of strain K as aB.

First, the data indicate the region homologous to probe A is linked to that homologous to probe B. This is reflected in the large excess of parental ditype asci (50) over non-parental ditypes (2). Therefore we need to map only one centromere.

The data show that the region homologous to probe A is on the opposite side of the centromere from the region homologous to probe B. This is reflected in the fact that the two single exchange classes (32 asci and 12 asci) show first division segregation for the A region only (12) or for the B region only (32), and you get second division segregation for both the A and B regions only in the double exchange asci (4, 2 and 2).

Considering the distance from region A to the centromere, we see that A second division segregation in 20 of the 100 asci. Gene-centromere distance is given by the formula: % second-division segregation asci/2, so here the gene-centromere distance is 20%/2 = 10 mu. To calculate this distance another way, the 20 asci contain 80 meiotic products, and half of these have had an exchange between A and the centromere. Thus 40 of a total of 400 meiotic products have experienced an exchange between A and the centromere. 40/400 = 0. 1, so the A region is 10 mu from the centromere.

The B region shows second division segregation in 40 asci, out of 100 assayed giving a gene-centromere distance of 40%/2 = 20 mu. Calculating gene-centromere distance another way, the 40 asci contain 80 meiotic products that have inherited chromosomes which have crossed over between B and the centromere. 80/400 = 0.2, so the B region is 20 mu from the centromere.

Given that A is 10 mu from the centromere, and B is 20 mu from the centromere, the one remaining fact to determine is whether A and B are on the same side of a centromere, or on opposite sides. If they were on the same side, the prediction would be that they are about 10 mu apart. If they were on the opposite sides of the opposite sides of the centromere, the map distance using tetrad analysis is given by the formula:

$$= \frac{\frac{1}{2}T + NPD}{Total \text{ \# Asci}} \times 100\%$$

$$= \frac{\frac{1}{2}(48) + 2}{100} \times 100\%$$

$$= 26 \text{ mu.}$$

Thus our map is:

$$A \xleftarrow{10}_{mu} \qquad \xleftarrow{\qquad 20 \qquad}_{mu} B$$

15.17   One appli____ ___n of DNA fingerprinting technology has been to identify stolen children and return them to their parents. Bobby Larson was taken from a supermarket parking lot in New Jersey in 1978, when he was 4 years old. In 1990, a sixteen-year-old boy called Ronald Scott was found in California, living with a couple named Susan and James Scott, who claimed to be his parents. Authorities suspected that Susan and James might be the kidnappers, and that Ronald Scott might be Bobby Larson. DNA samples were obtained from Mr. and Mrs. Larson, and from Ronald, Susan and James Scott. Then DNA fingerprinting was done, using a probe for a particular VNTR family, with the results shown in the following figure. From the information in the figure, what can you say about the parentage of Ronald Scott? Explain.

Answer: James and Susan Scott are not the parents of "Ronald Scott.". There are several bands in the fingerprint of the boy which are not present in either James or Susan, and thus could not have been inherited from either of them (note bands **a** and **b** in the answer figure, for example). In contrast, whenever the boy's DNA exhibits a band which is missing from one of member of the Larson couple, the other Larson has that band (for example, bands **c** and **d**). Thus there is no band in the boy's DNA that he could not have inherited from one or the other of the Larsons. These data indicate the boy is in fact Bobby Larson.

CHAPTER 16

REGULATION OF GENE EXPRESSION IN BACTERIA AND BACTERIOPHAGES

I. OUTLINE OF CHAPTER

Regulated and Constitutive Genes
Gene Regulation of Lactose Utilization in *E. coli*
    Lactose as a Carbon Source for *E. coli*
    Experimental Evidence for the Regulation of the *lac* Genes
    Jacob and Monod's Operon Model for the Regulation of the *lac* Genes
    Positive Control of the *lac* Operon
    Molecular Details of *lac* Operon Regulation
Tryptophan Operon of *E. coli*
    Gene Organization of the Tryptophan Biosynthesis Genes
    Regulation of the *trp* Operon
    Regulation of Other Amino Acid Biosynthesis Operons
Summary of Operon Function
Global Control of Transcription
Gene Regulation in Bacteriophages
    Regulation of Gene Expression in Phage T4
    Regulation of Gene Expression in Phage Lambda

II. IMPORTANT TERMS AND CONCEPTS

regulated genes	*lac* operon
constitutive genes	protein coding gene
inducers and effectors	catabolite repression
controlling site	tryptophan operon
induction	antitermination signal
coordinate induction	feedback inhibition
polygenic mRNA	allosteric shift
operator	stringent control
*lac*I	genetic switch
repressor molecule	CAP

III. ANALYTICAL THINKING

Owing to the complex and esoteric nature of the material presented in this chapter it is strongly recommended that especial attention be given to the explanatory diagrams. They should be studied diligently and reproduced, with each step in the particular sequence portrayed clearly indicated. When you can reproduce each diagram from memory, you may consider it learned and, it is hoped, understood.

IV. QUESTIONS FOR PRACTICE

A. Multiple Choice Questions

1. Genes in general that respond to the needs of a cell or organism in a controlled manner are known as
   a. inducer genes.
   b. effector genes.
   c. regulated genes.
   d. constitutive genes.

2. Genes that are always active in growing cells, irrespective of what the environmental conditions may be, are known as
   a. inducer genes.
   b. effector genes.
   c. regulated genes.
   d. constitutive genes.

3. A gene that is stimulated to undergo transcription in response to a particular molecular event that occurs at a controlling site near that gene is said to be
   a. inducible.
   b. constitutive.
   c. a promoter.
   d. an inducer.

4. β- galactosidase catalyzes the breakdown of lactose into
   a. 2 molecules of glucose.
   c. sucrose and galactose.
   b. 2 molecules of galactose
   d. galactose and glucose.

5. Lactose permease is an enzyme that is located in *E. coli* in
   a. response to lactose deficiency.
   b. the cell membrane.
   c. the *lac* operon.
   d. the operator site.

6. The actual inducer of the *lac* protein-coding genes is
   a. lactose.
   b. allolactose.
   c. glucose.
   d. β-galactosidase.

7. Bacterial operons of protein-coding genes for the synthesis of amino acids, such as the tryptophan operon, are customarily classified as
   a. negatively controlled.
   b. positively controlled.
   c. repressible operons.
   d. inducible operons.

8. Before transcription of the *lac* operon can occur, the RNA entry site must be destabilized. This happens when
   a. CAP binds to the CAP site in the promoter.
   b. CAP binds with cAMP.
   c. adenylcyclase converts ATP to cAMP.
   d. *lacI*$^+$ mutates to *lacI*$^-$.

9. Phage λ is maintained in the lysogenic state by a diffusible molecule, which is identified as
   a. Cro
   b. cI
   c. cII
   d. λ repressor

10. Integrated phage λ can be induced by ultraviolet light to enter the lytic pathway by
   a. cleaving repressor monomers.
   b. initiating transcription by the *cro* gene.
   c. converting protein RecA to a protease.
   d. all the above, directly or indirectly

Answers: 1c, 2d, 3a, 4d, 5b, 6b, 7c, 8a, 9d, 10d

B. Thought Questions

1.   What three proteins are synthesized when lactose is the sole carbon source in *E. coli*? What does each do? (See text pp. 478-480.)

2.   What are the roles of *lacA*, *lacI*, *lacO*, *lacY*, and *lacZ* in *E. coli* carbohydrate metabolism? In what order are they arranged in the DNA molecule? (See text p. 482 and Fig. 16.4.)

3.   Describe the sequence of events that occur in the lac genes in *E. coli* when grown in the presence of glucose and lactose. (See text pp. 487 ff.)

4.   Explain why the mutant gene *lacO^c*, is said to be *cis*-dominant. (See text pp. 483-484 and Fig. 16.7.)

5.   Compare the basic differences between the *lac* and the *trp* operons of *E. coli*.

V. ANSWERS AND SOLUTIONS TO TEXT QUESTIONS

16.1   How does lactose bring about the induction of synthesis of β-galactosidase, permease, and transacetylase? Why does this event not occur when glucose is also in the medium?

Answer: Lactose induces synthesis of the three enzymes as described on pp.482-483. The lactose causes a change in repressor shape, resulting in the repressor being released from the operator, thereby allowing initiation of transcription from the adjacent promoter.

   In the presence of glucose catabolite repression occurs, which prevents the transcription of catabolite-sensitive operons. A catabolite of glucose results in a decrease in the amount of cAMP in the cell, thereby reducing the amount of cAMP-CAP (catabolite activator protein) complex that must bind to promoters of catabolite-sensitive operons before RNA polymerase can bind.

16.2   Operons produce polygenic mRNA when they are active. What is a polygenic mRNA? What advantages, if any, do they confer on a cell in terms of its function?

Answer: Polygenic mRNA contains information for more than one protein, and since these mRNAs are transcribed from operons, the proteins are for related functions, such as catalyzing steps in a biochemical pathway. An advantage for polygenic mRNA is that it provides a convenient package for the coordinate production of proteins with related functions.

16.3   If an *E. coli* mutant strain synthesizes β-galactosidase whether or not the inducer is present, what genetic defect(s) might be responsible for this phenotype?

Answer: A constitutive phenotype can be the result of either a *lacI$^-$* or a *lacO^c* mutation.

16.4   Distinguish the effects you would expect from (a) a missense mutation and (b) a nonsense mutation in the *lacZ* (β-galactosidase) gene of the *lac* operon.

163

Answer:
a. A missense mutation results in partial or complete loss of β-galactosidase activity, but there would be no loss of permease and transacetylase activities.

b. A nonsense mutation is likely to have polar effects unless the mutation is very close to the normal chain-terminating codon for β-galactosidase. Thus permease and transacetylase activities would be lost in addition to the loss of β-galactosidase activity if nonsense mutation occurred near the 5' end of the *lacZ* gene.

16.5 The elucidation of the regulatory mechanisms associated with the enzymes of lactose utilization in *E. coli* was a landmark in our understanding of regulatory processes in microorganisms. In formulating the operon hypothesis as applied to the lactose system, Jacob and Monod found that results from particular partial-diploid strains were invaluable. Specifically, in terms of the operon hypothesis, what information did the partial diploids provide that the haploids could not?

Answer: Partial diploids were able to show (a) *cis-* and *trans-*dominant effects, (b) that repressor action was the result of a diffusible substance, (c) that operator function did not require a diffusible substance, and (d) that promoter function did not require a diffusible substance.

16.6 For the *E. coli lac* operon, write the partial-diploid genotype for a strain that will produce β-galactosidase constitutively and permease by induction.

Answer: $lacI^+$ $lacO^C$ $lacZ^+$ $lacY^-/lacI^+$ $lacO^+$ $lacZ^-$ $lacY^+$ (It cannot be ruled out that one of the repressor genes is $lacI^-$)

16.7 Mutants were instrumental in the elaboration of the model for the regulation of the lactose operon.
a. Discuss why $lacO^C$ mutants are cis-dominant but not trans-dominant.
b. Explain why $lacI^S$ mutants are trans-dominant to the wild-type $lacI^+$ allele but $lacI^-$ mutants are recessive.
c. Discuss the consequences of mutations in the repressor gene promoter as compared with mutations in the structural gene promoter.

Answer:
a. The $lacO^C$ mutants have an altered base pair in the operator that prevents the repressor from recognizing the operator. As a result, the structural genes cis (adjacent) to the $lacO^C$ are transcribed constitutively. The $lacO^C$ has no effect on other lactose operons in the same cell because $lacO^C$ codes for no product that could diffuse through the cell and affect other DNA sequences.

b. The $lacI^S$ mutants produce superrepressor molecules that have lost their ability to recognize the inducer. As a result, superrepressor molecules bind to normal lactose operators in the cell, no matter where they are, and prevent the transcription of the structural genes. In this case transdominance is seen because the $lacI^S$ mutants produce a diffusible product. Even with normal repressor molecules (from a $lacI^+$ gene) in the cell, the superrepressor molecules are "stuck" on the operators. $lacI^-$ mutants are recessive for the following reasons: $lacI^-$ mutations cause a change in the repressor molecule such that it cannot bind to the operator. However, if a $lacI^+$ gene is present in a partial diploid with $lacI^-$, a diffusible, functional repressor will be made that can bind to the operator and block transcription.

c. Mutations in the repressor gene promoter result in either an increase or a decrease in the level of expression of the repressor gene. Such mutations are likely to have little effect on the control of expression of the *lac* operon structural genes since the repressor molecule itself is unaltered. (The only possible effect would be if the number of repressor molecules made were low enough to cause increased structural gene expression in the absence of inducer.) Promoter mutations for the structural genes can also increase or decrease the level of expression of the (induced) operon. Most known promoter mutations cause an almost complete loss of expression of the three genes.

164

16.8   This question involves the lactose operon of *E. coli*. Complete the following table, using + to indicate if the enzyme in question will be synthesized and − to indicate if the enzyme will not be synthesized.

Table 16.A

	Genotype	Inducer absent		Inducer present	
		β-galactosidase	Permease	β-galactosidase	Permease
a.	$lacI^+ lacP^+ lacO^+ lacZ^+ lacY^+$				
b.	$lacI^+ lacP^+ lacO^+ lacZ^- lacY^+$				
c.	$lacI^+ lacP^+ lacO^+ lacZ^+ lacY^-$				
d.	$lacI^- lacP^+ lacO^+ lacZ^+ lacY^+$				
e.	$lacI^S lacP^+ lacO^+ lacZ^+ lacY^+$				
f.	$lacI^+ lacP^+ lacO^C lacZ^+ lacY^+$				
g.	$lacP lacP^+ lacO^C lacZ^+ lacY^+$				
h.	$lacI^+ lacP^+ lacO^C lacZ^+ lacY^-$				
i.	$lacI^{-d} lacP^+ lacO^+ lacZ^+ lacY^+$				
j.	$lacI^- lacP^+ lacO^+ lacZ^+ lacY^+/$ $lacI^+ lacP^+ lacO^+ lacZ^- lacY^-$				
k.	$lacI^- lacP^+ lacO^+ lacZ^+ lacY^-/$ $lacI^+ lacP^+ lacO^+ lacZ^- lacY^+$				
l.	$lacI^S lacP^+ lacO^+ lacZ^+ lacY^-/$ $lacI^+ lacP^+ lacO^+ lacZ^- lacY^+$				
m.	$lacI^+ lacP^+ lacO^C lacZ^- lacY^+/$ $lacI^+ lacP^+ lacO^+ lacZ^+ lacY^-$				
n.	$lacI^{-d} lacP^+ lacO^C lacZ^+ lacY^-/$ $lacI^+ lacP^+ lacO^+ lacZ^- lacY^+$				
o.	$lacI^S lacP^+ lacO^+ lacZ^+ lacY^+ /$ $lacI^+ lacP^+ lacO^C lacZ^+ lacY^+$				
p.	$lacI^{-d} lacP^+ lacO^+ lacZ^+ lacY^-/$ $lacI^+ lacP^+ lacO^+ lacZ^- lacY^+$				
q.	$lacI^+ lacP^- lacO^C lacZ^+ lacY^-/$ $lacI^+ lacO^+ lacO^+ lacZ^- lacY^+$				
r.	$lacI^+ lacP^- lacO^+ lacZ^+ lacY^-/$ $lacI^+ lacP^+ lacO^C lacZ^- lacZ^+$				
s.	$lacI^- lacP^- lacO^+ lacZ^+ lacY^+ /$				
t.	$lacI^- lacP^+ lacO^+ lacZ^+ lacY^-/$ $lacI^+ lacP^- lacO^+ lacZ^- lacY^+$				

Answer:

Genotype	Inducer absent β-galactosidase	Inducer absent Permease	Inducer present β-galactosidase	Inducer present Permease
a. $lacI^+ lacP^+ lacO^+ lacZ^+ lacY^+$	–	–	+	+
b. $lacI^+ lacP^+ lacO^+ lacZ^- lacY^+$	–	–	–	+
c. $lacI^+ lacP^+ lacO^+ lacZ^+ lacY^-$	-	-	+	-
d. $lacI^- lacP^+ lacO^+ lacZ^+ lacY^+$	+	+	+	+
e. $lacI^S lacP^+ lacO^+ lacZ^+ lacY^+$	–	–	–	–
f. $lacI^+ lacP^+ lacO^C lacZ^+ lacY^+$	+	+	+	+
g. $lacP lacP^+ lacO^C lacZ^+ lacY^+$	+	+	+	+
h. $lacI^+ lacP^+ lacO^C lacZ^+ lacY^-$	+	–	+	–
i. $lacI^{-d} lacP^+ lacO^+ lacZ^+ lacY^+$	+	+	+	+
j. $lacI^- lacP^+ lacO^+ lacZ^+ lacY^+/$ $lacI^+ lacP^+ lacO^+ lacZ^- lacY^-$	–	–	+	+
k. $lacI^- lacP^+ lacO^+ lacZ^+ lacY^-/$ $lacI^+ lacP^+ lacO^+ lacZ^- lacY^+$	–	–	+	+
l. $lacI^S lacP^+ lacO^+ lacZ^+ lacY^-/$ $lacI^+ lacP^+ lacO^+ lacZ^- lacY^+$	–	–	–	–
m. $lacI^+ lacP^+ lacO^C lacZ^- lacY^+/$ $lacI^+ lacP^+ lacO^+ lacZ^+ lacY^-$	–	+	+	+
n. $lacI^{-d} lacP^+ lacO^C lacZ^+ lacY^-/$ $lacI^+ lacP^+ lacO^+ lacZ^- lacY^+$	+	–	+	+
o. $lacI^S lacP^+ lacO^+ lacZ^+ lacY^+/$ $lacI^+ lacP^+ lacO^C lacZ^+ lacY^+$	+	+	+	+
p. $lacI^{-d} lacP^+ lacO^+ lacZ^+ lacY^-/$ $lacI^+ lacP^+ lacO^+ lacZ^- lacY^+$	+	+	+	+
q. $lacI^+ lacP^- lacO^C lacZ^+ lacY^-/$ $lacI^+ lacO^+ lacO^+ lacZ^- lacY^+$	–	–	–	+
r. $lacI^+ lacP^- lacO^+ lacZ^+ lacY^-/$ $lacI^+ lacP^+ lacO^C lacZ^- lacZ^+$	–	+	–	+
s. $lacI^- lacP^- lacO^+ lacZ^+ lacY^+/$ $lacI^+ lacP^+ lacO^+ lacZ^- lacY^-$	–	–	–	–
t. $lacI^- lacP^+ lacO^+ lacZ^+ lacY^-/$ $lacI^+ lacP^- lacO^+ lacZ^- lacY^+$	–	–	+	–

16.9 A new sugar, sugarose, induces the synthesis of two enzymes from the *sug* operon of *E. coli.* Some properties of deletion mutations affecting the appearance of these enzymes are as follows (here, + = enzyme induced normally, i.e., synthesized only in the presence of the inducer, C = enzyme synthesized constitutive; 0 = enzyme cannot be detected):

Mutation of	Enzyme 1	Enzyme 2
Gene A	+	0
Gene B	0	+
Gene C	0	0
Gene D	C	C

166

a. The genes are adjacent in the order *ABCD*. Which gene is most likely to be the structural gene for enzyme 1?

b. Complementation studies using partial-diploid (*F'*) strains were made. The episome (*F'*) and chromosome each carried one set of *sug* genes. The results were as follows (symbols are the same as in the previous table):

Genotype of *F'*	Chromosome	Enzyme 1	2
$A^+ B^- C^+ D^+$	$A^- B^+ C^+ D^+$	+	+
$A^+ B^- C^- D^+$	$A^- B^+ C^+ D^+$	+	0
$A^- B^+ C^- D^+$	$A^+ B^- C^+ D^+$	0	+
$A^- B^+ C^+ D^+$	$A^+ B^- C^+ D^-$	+	+

From all the evidence given, determine whether the following statements are true or false:
(1) It is possible that gene *D* is a structural gene for one of the two enzymes.
(2) It is possible that gene *D* produces a repressor.
(3) It is possible that gene *D* produces a cytoplasmic product required to induce genes *A* and *B*.
(4) It is possible that gene *D* is an operator locus for the *sug* operon.
(5) The evidence is also consistent with the possibility that gene *C* could be a gene that produces a cytoplasmic product *required* to induce genes *A* and *B*.
(6) The evidence is also consistent with the possibility that gene *C* could be the controlling end of the *sug* operon (end from which mRNA synthesis presumably commences).

Answer:
a. Gene *B*; only a mutation in gene *B* produced a loss of enzyme 1 activity with no effect on enzyme 2 activity.

b. (1) False; mutation in *D* leads to constitutive synthesis of *both* enzymes 1 and 2, so *D* cannot be a structural gene for one of the enzymes.

(2) True; *D* could code for a repressor. If the repressor acted like the *lac* repressor, then *D* mutants would lose the ability for the repressor to bind to the operator, and the structural genes in the operon would be expressed constitutively; D^- mutants on this model should be recessive to D^+; *D* mutants show these phenotypes.

(3) False; if *D* was needed to induce the *sug* operon, then D^- mutants should produce no enzymes, a result that was not observed.

(4) False; if *D* was the operator, then D^- mutants would be constitutive. However, operator mutants are cis-dominant and show no effect of the presence of a normal operator also in the cell. Since D^+ is trans-dominant to D^-, *D* cannot be an operator.

(5) False; C^- mutants show no production of enzymes 1 and 2. The C^- mutants have cis-dominant effects according to partial-diploid data; that is, $A^+ B^- C^- D^+/A^- B^+ C^+ D^+$ shows no production of enzyme 2, which, following logic similar to that in part a, is the product of structural gene *A*, and $A^- B^+ C^- D^+/A^+ B^- C^* D^+$ shows no production of enzyme 1, the product of structural gene *B*. In both cases the gene not expressed is the gene cis to the C^- locus.

(6) True; the cis-dominant effects of C^- mutants can be explained by the mutations being in the controlling end of the *sug* operon, such as in the promoter.

16.10 Four different polar mutations, *1, 2, 3,* and *4*, in the *lacZ* gene of the lactose operon were isolated following mutagenesis of *E. coli*. Each caused total loss of β-galactosidase activity. Two revertant mutants, due to suppressor mutation in genes unlinked to the *lac* operon, were isolated from each of the four strains. Suppressor mutations of polar mutation *1* or *1A* and *1B*; those of polar mutation *2* or *2A* or *2B*, and so on. Each of the eight suppressor mutations was then tested, by appropriate crosses, for its ability to suppress each of the four polar mutations; the test involved examining the ability of a strain carrying the polar mutation and the suppressor mutation to grow with lactose as the sole carbon source. The results follow (+ = growth on lactose and − = no growth):

| Polar | Suppressor Mutation | | | | | | | |
Mutation	1A	1B	2A	2B	3A	3B	4A	4B
1	+	+	+	+	+	+	+	+
2	+	−	+	+	+	+	−	−
3	+	−	+	−	+	+	−	−
4	+	+	+	+	+	+	+	+

A mutation to a UAG codon is called an amber nonsense mutation, and a mutation to a UAA codon is called an ochre nonsense mutation. Suppressor mutations allowing reading of UAG and UAA are called amber and ochre suppressors, respectively.

- a. Which of the polar mutations are probably amber? Which are probably ochre?
- b. Which of the suppressor mutations are probably amber suppressors? Which are probably ochre suppressors?
- c. How would you explain the anomalous failure of suppressor 2B to permit growth with polar mutation 3? How could you test your explanation most easily?
- d. Explain precisely why ochre suppressors suppress amber mutants but amber suppressors do not suppress ochre mutants.

Answer:

- a. Ochre suppressors suppress both ochre and amber mutants, but amber suppressors suppress only amber mutants. Thus amber mutants will be those suppressible by all given suppressors, that is, 1 and 4. Since not all suppressors suppress 2 and 3, these mutations are probably ochre mutants.

- b. Since an ochre suppressor will allow growth of all four polar mutations, and an amber suppressor will allow growth of only 1 and 4, then 1B, 4A, and 4B are amber suppressors, and 1A, 2A, 3A, and 3B are ochre suppressors (2B will be discussed in part c).

- c. Since 2B suppresses 1, 2, and 4, it is probably an ochre suppressor. The reason 3 is not suppressed is explicable in terms of the suppression mechanism; that is, it involves synthesis of a tRNA molecule with an anticodon to ochre triple UAA. This tRNA will carry an amino acid probably different from that coded for by the original triplet. Thus an incorrect amino acid is inserted into the B-galactosidase protein. If this amino acid caused the protein to be no longer functional, the enzyme could not break down lactose and the mutant would not grow (i.e., would not be suppressed) on lactose. By this hypothesis a protein is produced, but it is nonfunctional. One can test for this result since it is cross-reacting material (i.e., has a structure similar to that of a functional enzyme, so is precipitable by antibodies made against the wild-type enzyme). So by use of antibodies produced against the wild-type enzyme, any similar reaction indicates a related substance present (i.e., cross-reacting material for β-galactosidase), and this result would be evidence for a protein being produced that is nonfunctional. If the test works for 2B, the test would explain the result.

- d. This result may be explained by the wobble hypothesis, which says that there is some flexibility in the base pairing between the tRNA anticodon and the mRNA triplet at the third nucleotide. Thus U (uracil) in the tRNA has been found to be able to bind both with A or G in the mRNA. Now the ochre codon is UAA and the anticodon is UUA (in tRNA). The third base U in the anticodon can pair with A (ochre triplet) or G (amber triplet UAG) by the wobble hypothesis. Since ochre suppressors have a tRNA species with such an anticodon, and since there is wobble, ochres can suppress amber as well as ochre mutants. The reciprocal is not true, however. The amber triplet is UAG, and anticodon in the suppressor tRNA is CUA. Now the third base C has no wobble properties and can only pair with its normal partner G. That is, the only codon that the anticodon CUA will fit is the amber codon UAG, whereas the ochre suppressor tRNA anticodon UUA can pair with both UAA and UAG (efficiency of suppression is lower for UAG) because of wobble of U for A or G.

16.11 What consequences would a mutation in the catabolite activator protein (CAP) gene of *E. coli* have for the expression of a wild-type *lac* operon?

Answer: The CAP, in a complex with cAMP, is required to facilitate RNA polymerase binding to the *lac* promoter. The RNA polymerase binding occurs only in the absence of glucose and only if the operator is not occupied by repressor (i.e., if lactose is absent). A mutation in the CAP gene, then, would render the lac operon incapable of expression since RNA polymerase would, not be able to recognize the promoter.

16.12 The lactose operon is an inducible operon, whereas the tryptophan operon is a repressible operon. Discuss the differences between these two types of operons.

Answer: Inducible and repressible operons are similar; they differ in the details of the control of transcription. For example, in the lactose operon the system is off when lactose is absent, and it is on when lactose is present. Mechanistically, this result is brought about by the fact that the lactose operon repressor can bind to the operator in the absence of lactose, but when lactose is added, lactose binds to the repressor, preventing it from binding to the operator. The operon is then transcribed. In a repressible operon, such as one that codes for the enzymes that catalyze the steps in a biochemical pathway (e.g., for tryptophan biosynthesis), the strategy is the opposite. When the amino acid is present in the medium, the operon should be turned off. In this situation, the amino acid binds to a corepressor protein and changes its shape so that it can bind to the operator and block transcription. In the absence of the amino acid the corepressor has no affinity for the operator, and the operon is active.

16.13 In the bacterium *Salmonella typhimurium* seven of the genes coding for histidine biosynthetic enzymes are located adjacent to one another in the chromosome. If excess histidine is present in the medium, the synthesis in all seven enzymes is coordinately repressed, whereas in the absence of histidine all seven genes are coordinately expressed. Most mutations in this region of the chromosome result in the loss of activity of only one of the enzymes. However, mutations mapping to one end of the gene cluster result in the loss of all seven enzymes, even though none of the structural genes have been lost. What is the counterpart of these mutations in the *lac* operon system?

Answer: lacP⁻ (promoter) mutations

16.14 What is stringent control, and how does this regulatory system work?

Answer: See text pp. 498-499. Stringent control is the regulatory mechanism at bacteria to stop essential cellular biochemical activities when the cell is in a harsh nutritional condition, such as when a carbon source runs out or when an auxotroph is starved for the nutrient involved. Characteristically, the unusual nucleotides ppGpp and pppGpp accumulate rapidly when the stringent-control system is in effect.

16.15 Bacteriophage λ, upon infecting an *E. coli* cell, has a choice between the lytic and lysogenic pathways. Discuss the molecular events that determine which pathway is taken.

Answer: See text, pp. 499-507.

16.16 How do the lambda repressor protein and the Cro protein regulate their own synthesis?

Answer: See text, pp. 502-507.

16.17 If a mutation was induced in the *cI* gene of phage lambda such that the resulting *cI* gene product was nonfunctional, what phenotype would you expect the phage to exhibit?

Answer: The *cI* gene codes for repressor protein that functions to keep the lytic functions of the phage repressed when lambda is in the lysogenic state. A *cI* mutant strain would be unable to establish lysogeny, and thus it would also follow the lytic pathway.

16.18 Bacteriophage λ can form a stable association with the bacterial chromosome because the virus manufactures a repressor. This repressor prevents the virus from replicating its DNA, making lysozyme and all the other tools used to destroy the bacterium. When you induce the virus with UV light, you destroy the repressor, and the virus goes through its normal lytic cycle. This repressor is the product of a gene called the cI gene and is a part of the wild-type viral genome. A bacterium that is lysogenic for λ^+ is full of repressor substance, which confirms immunity against any λ virus added to these bacteria. These added viruses can inject their DNA, but the repressor from the resident virus prevents replication, presumably by binding to an operator on the incoming virus. Thus this system has many analogous elements to the lactose operon. We could diagram a virus as shown in the figure. Several mutations of the cI gene are known. The c_i mutation results in an inactive repressor.

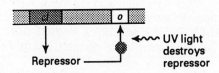

a. If you mix λ containing a c_i mutation, can it lysogenize (form a stable association with the bacterial chromosome)? Why?

b. If you infect a bacterium simultaneously with a wild-type c^+ and a c_i mutant of λ, can you obtain stable lysogeny? Why?

c. Another class of mutants called c^{IN} makes a repressor that is insensitive to UV destruction. Will you be able to induce a bacterium lysogenic for c^{IN} with UV light? Why?

Answer: a. No; the lambda genome cannot be repressed, and only the lytic pathway can operate in the presence of a c_i mutation.

b. Yes; a normal repressor is made from the wild-type repressor gene and would be transdominant to the c_i mutant.

c. No because the repressor is insensitive to UV light. The lambda genome is stuck in the prophage state.

CHAPTER 17

REGULATION OF GENE EXPRESSION AND DEVELOPMENT IN EUKARYOTES

I. CHAPTER OUTLINE

Levels of Control of Gene Expression in Eukaryotes
    Transcriptional Control
    RNA Processing Control
    Transport Control
    mRNA Translation Control
    mRNA Degradation Control
Gene Regulation in Development and Differentiation
    Gene Expression in Higher Eukaryotes
    Constancy of DNA in the Genome During Development
    Differential Gene Activity in Tissues and During Development
    Immunogenetics
Genetic Regulation of Development in *Drosophila*
    Drosophila Developmental Stages
    Embryonic Development
    Imaginal Discs
    Homeotic Genes
Genetic Regulation of Development in the Nematode *Caenorhabditis elegans*

II. IMPORTANT TERMS AND CONCEPTS

transcriptional control	degradation control
regulatory elements	development
combinatorial gene regulation	differentiation
erythroblast	totipotent
DNase hypersensitivity site	endoreduplication
DNA methylation	immunoglobulin
short term gene regulation	clonal selection
steroid response elements	somatic recombination
processing control	imaginal disc
transport control	transdetermination
spliceosome	homeotic gene
retention	homeotic mutation
processing control	homeobox
transport control	translational control

III. ANALYTICAL THINKING

Once again we are dealing with material that requires very close attention to detail, particularly the subtle differences between certain control mechanisms. Frequent reference to the explanatory figures and diagrams as a very significant part of the descriptive text is recommended. It is helpful to refer back to the analogous mechanisms of regulation of gene expression in bacteria and viruses for a better understanding of their counterpart in the eukaryotes, and to think about the reasons for the differences.

IV. QUESTIONS FOR PRACTICE

A. Multiple Choice Questions

1. Gene expression is commonly regulated by unified collections of protein-coding genes and adjacent controlling sites, constituting what is known as operons, in
   a. prokaryotes.
   b. eukaryotes.
   c. both a and b
   d. certain viruses only.

2. Gene expression in some organisms may be regulated by
   a. transcriptional control.
   b. transport control.
   c. RNA degradation control.
   d. any of the foregoing.

3. Transcription in eukaryotes is activated if positive regulatory proteins are bound at
   a. the enhancer element.
   b. the promoter element.
   c. the operator.
   d. both a and b

4. Histones act in eukaryotic gene regulation primarily as
   a. gene enhancers .
   b. gene repressors.
   c. promoters.
   d. both a and c

5. Sensitivity to digestion by DNase I is pronounced in genes that are
   a. bound to histones.
   b. bound to nucleosomes.
   c. transcriptionally active.
   d. under transport control.

6. The DNA of most eukaryotic cells has been shown to be extensively
   a. methylated.
   b. degraded.
   c. bound to nucleosomes.
   d. DNase I sensitive.

7. The receptors for steroid hormones are located
   a. inside the target cells.
   b. in the cell membranes of the target cells.
   c. on the surface of the target cells.
   d. at the cell/dendrite interface.

8. The steroid-receptor complex that forms between a steroid hormone and its unique receptor enters the nucleus and binds to a specific site on DNA termed a
   a. steroid activator site.
   b. steroid response element.
   c. transcriptional site.
   d. transcription activator.

9. Choice of poly(A) site is a mechanism that characterizes
   a. transcriptional control.
   b. RNA processing control.
   c. transport control.
   d. RNA degradation control.

10. During embryonic differentiation and development the nuclear DNA per cell shows
   a. a progressive shift in base-pair ratios.
   b. an increase in total base-pair number.
   c. a progressive diminution in amount.
   d. no change in amount.

Answers: 1a, 2d, 3d, 4b, 5c, 6a, 7a, 8b, 9b, 10d

B. Thought Questions

1. Design an experimental procedure for determining whether or not a given gene is available for transcription. (Hint: Use of a restriction enzyme and an endonuclease might figure in your design.)

2. Present evidence to support the hypothesis that the histones function as <u>gene</u> repressors in eukaryotic cells. (See text, pp. 518-520 ff.)

3. What role, if any, does DNA methylation play in transcriptional control in some eukaryotes? (See text pp. 520-522.)

4. Propose a hypothesis to explain why it is that steroid hormone receptors are located inside the target cells, whereas the receptors for proteinaceous hormones are located on the surfaces of their target cells. (Hint: What are the permeability characteristics of cell membranes?)

5. Describe and evaluate the results of John Gurdon's experiments with nuclear transplants in *Xenopus laevis*. (See text, pp. 533-534, Fig. 17.17.)

6. What are homeoboxes and what is their role in the regulation of development in eukaryotes? (See text, pp. 547-550.)

V. ANSWERS AND SOLUTIONS TO TEXT QUESTIONS

17.1   Eukaryotic organisms have a large number of copies (usually more than a hundred) of the genes that code for ribosomal RNA, yet they have only one copy of each gene that codes for each ribosomal protein. Explain why.

Answer: The final product of the rRNA genes is an rRNA molecule. Hence a large number of genes are required to produce the large number of rRNA molecules required for ribosome biosynthesis. Ribosomal proteins, in contrast, are the end products of the translation of mRNAs, which can be "read" over and over to produce the large number of ribosomal protein molecules required for ribosome biosynthesis.

17.2   The human α, β, γ δ, ε and ζ globin genes are transcriptionally active at various stages of development. Fill in the following table, indicating whether the globin gene in question is sensitive (S) or resistant (R) to DNase I digestion at the developmental stages listed.

Globin Gene	Tissue		
	Embryonic Yolk Sac	Spleen	Adult Bone Marrow
β			
γ			
δ			
ζ			
ε			

Answer:

Globin Gene	Tissue		
	Embryonic Yolk Sac	Spleen	Adult Bone Marrow
β	equal	less	equal
γ	equal	equal	less
δ	equal	less	equal
ζ	greater	less	less
ε	greater	less	less

17.3   A cloned DNA sequence was used to probe a Southern blot. There were two DNA samples on the blot, one from white blood cells and the other from a liver biopsy of the same individual. Both samples had been digested with *Hpa*II. The probe bound to a single 2.2 kb band in the white blood cell DNA, but bound to two bands (1.5 and 0.7 kb) in the liver DNA.
   a.   Is this difference likely to be due to a somatic mutation in a *Hpa*II site? Explain.
   b.   How would it affect your answer if you knew that white blood cell and liver DNA from this individual both showed the 2 band pattern when digested with *Msp*I?

Answer: a.   Going from a single 2.2 kb band to a 1.5 and a 0.7 kb band (or vice versa) would actually require two somatic mutations (heterozygous cells would have all three bands), so this is an unlikely explanation.
   b.   Both *Hpa*II and *Msp*I recognize the CCGG site, but *Msp*I will, and *Hpa*II will not, cut when the internal C is methylated. Thus the data fit the idea that a particular CCGG site in the region homologous to the probe was methylated in white blood cells but not in liver in this person.

17.4   What is a hormone?

Answer: A hormone is a chemical messenger transmitted in body fluids from one part of the organism to another. It produces a specific effect on target cells that may be remote from its point of origin, and it functions to regulate gene activity, physiology, growth, differentiation, or behavior.

17.5   How do hormones participate in the regulation of gene expression in eukaryotes?

Answer: See text, pp. 523-526. Some hormones exert their effects by binding directly to the cell's genome and causing changes in gene expression. Other hormones act at the cell surface to activate a system that produces cAMP. The cAMP then acts as a second messenger molecule to activate the cellular events normally associated with the hormone involved. Five classes of hormones play roles in controlling growth and development in plants (see text, pp. 526-528.)

17.6   The following figure shows the effect of the hormone estrogen on ovalbumin synthesis in the oviduct of 4-day-old chicks. Chicks were given daily injections of estrogen ("Primary Stimulation") and then after 10 days the injections were stopped. Two weeks after withdrawal (25 days), the infections were resumed ("Secondary Stimulation").

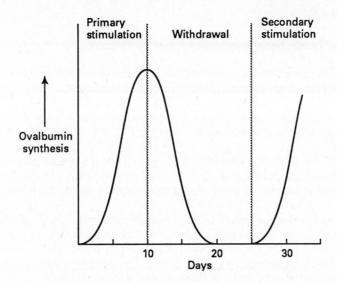

Provide possible explanations of these data.

Answer: The data show clearly that the synthesis of ovalbumin is dependent upon the presence of the hormone estrogen. The data do not indicate at what level estrogen works. Theoretically, it could act, to increase transcription of the ovalbumin gene, to, stabilize the ovalbumin mRNA, to stabilize the ovalbumin protein, or to stimulate transport of the ovalbumin mRNA out of the nucleus. Experiments in which the levels of the ovalbumin mRNA were measured have shown that the production of ovalbumin mRNA is primarily regulated at the level of transcription.

17.7   Distinguish between the terms *development* and *differentiation*.

Answer: In brief, development is a process of regulated growth and differentiation that results from the interaction of the genome with cytoplasm, internal cellular environment, and external environment. In multicellular organisms differentiation is the aspect of development that involves the formation of distinctly different types of cells and tissues from a zygote through processes that are specifically regulated by genes.

17.8   What is totipotency? Give an example of the evidence for the existence of this phenomenon.

Answer: Totipotency refers to the capacity of a nucleus to direct a cell through all the stages of development. In other words, a cell taken from a differentiated tissue is totipotent if it can be isolated and if a complete functional organism can develop from it. The implication is that the cell contains all the genetic information present in a zygote so that the developmental program for the complete organism can be executed. The classic demonstration of totipotency was done by Gurdon with *Xenopus laevis* and is described in detail in the text pp. 533-534.

17.9   Discuss some of the evidence for differential gene activity during development.

Answer: See text, pp. 535-538. Your answer should include discussion of lactate dehydrogenase isozymes, expression of human globin genes at different stages of development, and polytene chromosome puffing patterns.

17.10   The enzyme lactate dehydrogenase (LDH) consists of four polypeptides (a tetramer). Two genes are known to specify two polypeptides, A and B, which combine in all possible ways (A_4, A_3B, A_2B_2, AB_3, and B_4) to produce five LDH isozymes. If, instead, LDH consisted of three polypeptides (i.e., it was a trimer), how many possible isozymes would be produced by various combinations of polypeptides A and B?

Answer: four: A_3, A_2B, AB_2, B_3

17.11   Discuss the expression of human hemoglobin genes during development.

Answer: See text, pp. 535-536. Your answer should include a discussion of the various α-like and β-like globin genes and their expression at different times of development.

17.12   Discuss the organization of the hemoglobin genes in the human genome. Is there any correlation with the temporal expression of the genes during development?

Answer: See text, pp. 535-536. Your answer should include the fact that the α–like genes are clustered on one chromosome and the β-like genes are clustered on a different chromosome. Each of these two gene clusters is arranged in order of its temporal expression during development.

17.13   In humans, β-thalassemia is a disease caused by failure to produce sufficient β-globin chains. In many cases, the mutation causing the disease is a deletion of all or part of the β-globin structural gene. Individuals homozygous for certain of the β-thalassemia mutations are able to survive because their bone marrow cells produce γ-globin chains. The γ-globin chains combine with α-globin chains to produce fetal hemoglobin. In these people, fetal hemoglobin is produced by the bone marrow cells throughout life, whereas normally it is produced in the fetal liver. Use your knowledge about gene regulation during development to suggest a mechanism by which this expression of γ-globin might occur in β-thalassemia.

Answer: There are various possibilities. One fairly simple one would be to imagine that the γ-globin is normally under negative regulation by β-globin in bone marrow, and that when β-globin is not formed the γ-globin gene is derepressed.

17.14   What are polytene chromosomes? Discuss the molecular nature of the puffs that occur in polytene chromosomes during development.

Answer: See text, pp. 536-539. The puffs in polytene chromosomes are a visual representation of differential gene gene activity. The puffs are regions where the multiple copies of the chromosomes have loosened their tightly packed arrangement of DNA to expose sites for transcription. Puffs are active regions of RNA synthesis.

17.15   Puffs of regions of the polytene chromosomes in salivary glands of *Drosophila* are surrounded by RNA molecules. How would you show that this RNA is single-stranded and not double-stranded?

Answer: If the RNA is double-stranded, it can be heat-denatured into two single strands, which may differ in GC content and could be separated by cesium chloride equilibrium density gradient centrifugation. If the RNA is double-stranded and it is radioactively labeled, it will not be able to hybridize with denatured (single-stranded) genomic DNA.

17.16   In experiment A, ^{3}H-thymidine (a radioactive precursor of DNA) is injected into larvae of *Chironomus*, and the polytene chromosomes of the salivary glands are later examined by autoradiography. The radioactivity is seen to be distributed evenly throughout the polytene chromosomes. In experiment B, ^{3}H-uridine (a radioactive precursor of RNA) is injected into the larvae, and the polytene chromosomes are examined. The radioactivity is first found only around puffs; later, radioactivity is also found in the cytoplasm. In experiment C, actinomycin D (an inhibitor of transcription) is injected into larvae and then ^{3}H-uridine is injected. No radioactivity is found associated with the polytene chromosomes, and few puffs are seen. Those puffs that are present are much smaller than the puffs found in experiments A and B. Interpret these results.

Answer: Experiment A results in all the DNA becoming radioactively labeled, and so since DNA is a fundamental and major component of polytene chromosomes, radioactivity is evident throughout the chromosomes. Experiment B results in radioactive labeling of RNA molecules. Since radioactivity is first found only around puffs and later in the cytoplasm, we can hypothesize that the puffs are sites of transcriptional activity. Initially, radioactivity is found in RNA that is in the process of being synthesized, and the later appearance of radioactivity in the cytoplasm reflects the completed RNA molecules that have left the puffs and are being translated in the cytoplasm. Experiment C provides additional support for the hypothesis that transcriptional activity is-associated with puffs. That is, the inhibition of RNA transcription by actinomycin D blocks the appearance of ^{3}H-uridine (which would be in RNA) at puffs. In fact, puffs are much smaller, indicating that the puffing process is intimately associated with the onset of transcriptional activity for the gene(s) in that region of the chromosomes.

17.17   The following figure shows the percentage of ribosomes found in polysomes in unfertilized sea urchin oocytes (0 h) and at various times after fertilization:

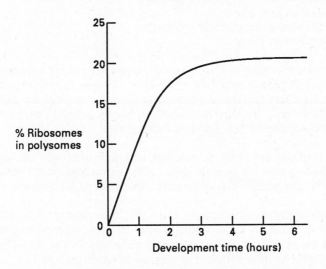

In the unfertilized egg, less than 1% of ribosomes are present in polysomes, while at 2 h post-fertilization, about 20% of ribosomes are present in polysomes. It is known that no new mRNA is made during the time period shown. How may the data be interpreted?

Answer: Pre-existing mRNAs (stored in the oocyte) are recruited into polysomes as development begins following fertilization.

17.18   The mammalian genome contains about 10^5 genes. Mammals can produce about 10^6 to 10^8 different antibodies. Explain how it is possible for both of the above sentences to be true.

Answer: Although we can make 10^6 to 10^8 different antibodies, we do not have a separate gene for each of these antibodies. Instead, our genomes encode several different forms of light and heavy chain constant regions, and a variety of variable and joining region sequences. An enormous array of antibody genes is then generated in B cells by somatic recombination such that a given variable region is attached to a given joining region and a particular constant region. The possible number of combinations of these parts is very large.

17.19   Define *imaginal disc, homeotic mutant,* and *transdetermination.*

Answer: See text, pp. 545-547.

17.20 Imagine that you observed the following mutants (a through e) in *Drosophila*. Based on the characteristics given, assign each of the mutants to one of the following categories: maternal gene, segmentation gene, or homeotic gene.

    a. Mutant *a:* In homozygotes phenotype is normal, except wings are oriented backwards.

    b. Mutant *b*: Homozygous females are normal but produce larvae that have a head at each end and no distal ends. Homozygous males produce normal offspring (assuming the mate is not a homozygous female).

    c. Mutant *c*: Homozygotes have very short abdomens, which are missing segments AB2 through AB4.

    d. Mutant *d*: Affected flies have wings growing out of their heads in place of eyes.

    e. Mutant *e*: Homozygotes have shortened thoracic regions and lack the second and third pair of legs.

Answer:

Mutant	Class
a	segmentation gene (segment polarity)
b	maternal gene (anterior-posterior gradient)
c	segmentation gene (gap)
d	homeotic (eye to wing transdetermination)
e	segmentation gene (gap)

17.21 If actinomycin D, an antibiotic that inhibits RNA synthesis, is added to newly fertilized frog eggs, there is no significant effect on protein synthesis in the eggs. Similar experiments have shown that actinomycin D has little effect on protein synthesis in embryos up until the gastrula stage. After the gastrula stage, however, protein synthesis is significantly inhibited by actinomycin D, and the embryo does not develop any further. Interpret these results.

Answer: Preexisting mRNA that was made by the mother and deposited into the egg prior to fertilization is translated up until the gastrula stage. After gastrulation new mRNA synthesis is necessary for production of proteins needed for subsequent embryo development.

17.22 It is possible to excise small pieces of early embryos of the frog, transplant them to older embryos, and follow the course of development of the transplanted material as the older embryo develops. A piece of tissue is excised from a region of the late blastula or early gastrula that would later develop into an eye and is transplanted to three different regions of an older embryo host (see a in the following figure). If the tissue is transplanted to the head region of the host, it will form eye, brain, and other material characteristic of the head region. If the tissue is transplanted to other regions of the host, it will form organs and tissues characteristic of those regions in normal development (e.g., ear, kidney, etc.). In contrast, if tissue destined to be an eye is excised from a neurula and transplanted into an older embryo host to exactly the same places as used for the blastula/gastrula transplants, in every case the transplanted tissue differentiates into an eye (see b in the following figure). Explain these results.

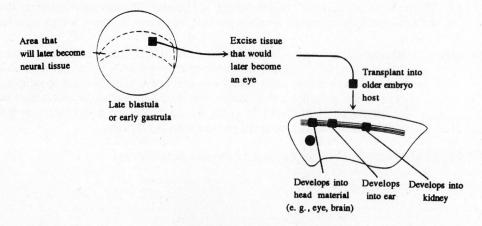

(a) Tissue from late blastula or early gastrula

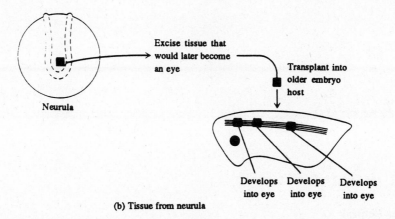

Neurula

Excise tissue that
would later become
an eye

Transplant into
older embryo
host

Develops
into eye

Develops
into eye

Develops
into eye

(b) Tissue from neurula

Answer: The tissue taken from the blastula/gastrula has not yet been committed to its final differentiated state in terms of its genetic programming; that is, it is not yet determined. Thus when it is transplanted into the host, the determined tissues surrounding the transplant in the host communicate with the transplanted tissue and cause it to be determined in the same way as they are; for example, tissue transplanted to a future head area will differentiate into head material, and so on. in contrast, tissues in the neurula stage are now determined as to their final tissue state once development is complete. Thus tissue transplanted from a neurula to an older embryo cannot be influenced by the determined, surrounding tissues and will develop into the tissue type for which it is determined, in this case an eye.

CHAPTER 18

GENE MUTATIONS

I. CHAPTER OUTLINE

Gene Mutations Defined
    Types of Mutations
    Reverse Mutations and Suppressor Mutations
Causes of Mutations
    Spontaneous Mutations
    Induced Mutations
The Ames Test: A Screen for Potential Mutagens and Carcinogens
DNA Repair Mechanisms
    Direct Correction of Mutational Lesions
    Repair Involving Excision of Base Pairs
    Human Genetic Diseases Resulting from DNA Replication and Repair Errors
Screening Procedures for the Isolation of Mutants
    Visible Mutations
    Nutritional Mutations
    Conditional Mutations

II. IMPORTANT TERMS AND CONCEPTS

mutation	mutation rate
chromosomal aberration	mutation frequency
genome mutation	tautomer
gene mutation	tautomeric shift
point mutation	SOS system
mutagen	base analog
induced mutation	base modifying agent
spontaneous mutation	intercalating agent
transition mutation	Ames test
transversion mutation	reverse genetics
missense mutation	photoreactivation
nonsense mutation	mutator mutation
neutral mutation	proofreading
silent mutation	alkylating repair
frameshift mutation	excision (dark) repair
reverse mutation	glycosylase
suppressor mutation	mismatch correction
intragenic suppressor	somatic mutation
intergenic suppressor	germline mutation
suppressor gene	conditional mutation

III. ANALYTICAL THINKING

There are many new terms and definitions in this chapter, and in order to follow some of the complicated processes described, it is essential that one have the meaning of each term well in mind. When dealing with some of the processes and some of the problems, it helps to go back to the DNA and diagram the sequence of base-pairs before and after each of the different classes of point mutations, followed, where indicated, by the base sequence of the mRNA transcribed from the mutant gene in question. The mechanisms of gene repression deserve special attention, and once again diagramming examples of them will pay dividends in understanding.

IV. QUESTIONS FOR PRACTICE

A. Multiple Choice Questions

(Use the following list of choices for Questions 1-6)
a. missense mutation
b. transversion mutation
c. neutral mutation
d. forward mutation

e. nonsense mutation
f. transition mutation
g. frameshift mutation
h. suppressor mutation

1. A mutation that involves a change in DNA at a particular site from one purine-pyrimidine base pair to a different purine-pyrimidine base pair.

2. A base-pair alteration that changes a codon in the mRNA such that the resulting amino acid substitution produces no change in the function of the protein translated from that message.

3. A DNA change in which a purine-pyrimidine base-pair is substituted by a pyrimidine-purine base-pair at the same site.

4. A mutation resulting from the addition or deletion of a base-pair in a gene.

5. Any point mutation that is expressed as a change in phenotype from wild type to mutant.

6. A base-pair change that results in the change of a mRNA codon to either UAG, UAA, or UGA.

7. The chemical compounds, 5BU and 2AP are
   a. rare purine isomers.
   b. base analog mutagens.
   c. components of SOS genes.
   d. both a and b

8. The Ames Test is used to screen for
   a. potential mutagens and carcinogens.
   b. the presence of enol forms of thymine and guanine.
   c. frameshift mutations.
   d. spontaneous mutations.

9. A significant effect of UV-radiation is the
   a. production of base tautomers.
   b. breakage of phosphodiester bonds.
   c. formation of pyrimidine dimers.
   d. induction of auxotrophic mutants.

10.   Azidothymidine (AZT), an approved drug for AIDS patients, is a(n)

      a.   retrovirus.
      b.   base analog.
      c.   reverse transcriptase.
      d.   intercalating agent.

Answers: 1a, 2c, 3b, 4g, 5d, 6e, 7b, 8a, 9c, 10b

B.   Thought Questions

1.   Describe site-specific *in vitro* mutagenesis and discuss its usefulness as a research technique. (See text, pp. 572-573.)
2.   Describe the Ames Test and comment on its merits and limitations. (See text pp. 573-575.)
3.   Compare DNA proofreading and repair in eukaryotes and prokaryotes. (See text pp. 575-578.)
4.   Describe the technique of replica plating. (See text, pp. 580-581.)

V.   ANSWERS AND SOLUTIONS TO TEXT QUESTIONS

18.1   Mutations are (choose the correct answer):
       a.   Caused by genetic recombination.
       b.   Heritable changes in genetic information.
       c.   Caused by faulty transcription of the genetic code.
       d.   Usually but not always beneficial to the development of the individuals in which they occur.

Answer: b

1 8.2   Answer TRUE or FALSE: Mutations occur more frequently if there is a need for them.

Answer: False. Mutations are not to be confused with selection. Once they occur, mutations may be selected for or against by processes of natural (or artificial) selection, according to the "need" for them or the advantage or disadvantage of them. There is no substantial evidence of any correlation between frequency of occurrence of mutations and a need for them.

18.3   The following is not a class of mutation (choose the correct answer):
       a.   Frameshift.
       b.   Missense.
       c.   Tansition.
       d.   Tansversion.
       e.   None of the above (i.e., all are classes of mutation).

Answer: e. none of the above (i.e. all are classes of mutations)

18.4   Ultraviolet light usually causes mutations by a mechanism involving (choose the correct answer):
       a.   One-strand breakage in DNA.
       b.   Light-induced change of thymine in alkylated guanine.
       c.   Induction of thymine dimers and their persistence or imperfect repair.
       d.   Inversion of DNA segments.
       e.   Deletion of DNA segments.
       f.   All of the above.

Answer: c. induction of thymine dimers and their persistence or imperfect repair. The key to this answer is the word "usually". The other choices might apply as exceptionally rare occurrences, but certainly not usually.

18.5   For the middle region of a particular polypeptide chain, the normal amino acid sequence and the amino acid sequence of several mutants were determined, as shown in the following (.... indicates

182

additional, unspecified amino acids). For each mutant, say what DNA level change has occurred, and whether the change is a point mutation (transversion or transition, missense or nonsense) or a frame shift. (Refer to the codon dictionary in Figure 14.3.)

    a.   Normal: ....Phe Leu Pro Thr Val Thr Thr Arg Trp
    b.   Mutant 1: ....Phe Leu His His Gly Asp Asp Thr Val
    c.   Mutant 2: ....Phe Leu Pro Thr Met Thr Thr Arg Trp
    d.   Mutant 3: ....Phe Leu Pro Thr Val Thr Thr Arg
    e.   Mutant 4: ....Phe Pro Pro Arg
    f.   Mutant 5: ....Phe Leu Pro Ser Val Thr Thr Arg Trp

Answer: Mutant 1 is evidently a frameshift, since the entire amino acid sequence is altered after a given point. The normal first three codons (for Phe Leu Pro) must be UUU (or UUC) UUA or UUG or CUX (X = any base) and CCX, whereas the new third amino acid is His (CAU or CAC). We can generate CAU or CAC from CCX by inserting an A between the two Cs. The new fourth codon should then receive the leftover X plus AC from the normal fourth Thr codon. If X is C, this would give CAC (His). The leftover X from Thr plus GU from the fifth (Val) codon would give Gly (GGU) if X is G, and so on.

    The only difference between mutant 2 and normal is that the fifth amino acid is Met instead of Val, so this is likely to be a point mutation. AUG is the only codon for Met. Analysis of mutant 1 showed that the Val codon is GUG, and we can get AUG by a single base pair substitution. This is a missense mutation, and since the DNA change is CG to TA, it is a transition.

    Mutant 3 has a normal amino acid sequence, but is prematurely terminated, suggesting a nonsense mutation. Termination occurs where Trp is normally located. The Trp codon is UGG. Either UGA or UAG would be a stop codon. Evidently we have a point mutation affecting the second or third base pair. In either case the DNA change is CG to TA, which is a transition.

    Mutant 4 also shows premature termination, but in this case the preceding amino acid sequence is not normal, suggesting that the basic change is a frameshift mutation. We have established that the third, fourth and fifth codons are CCC ACG GUG. If we try to make a frameshift by inserting a base pair into the first or second codon, so that we get CCU (Pro) instead of CUX (Leu), the third codon will still give Pro (as it should) if X is C. However, then the fourth codon would encode His (not Arg), and indeed the amino acid sequence after this point should be as in mutant 1. If we instead make a deletion of one base pair in the DNA corresponding to the second codon, we can explain mutant 4. If Leu comes from CUC and we delete the U, we would get CCC (Pro). The third codon would have its remaining CC, and would get A from the fourth codon, making CCA (Pro). CG would remain in the fourth codon, and G would come in from the fifth, making CGG (Arg). UG would remain from the fifth codon, and A would come in from the sixth, making UGA, a stop codon.

    Mutant 5 shows an alteration of only the fourth amino acid (Thr to Ser), suggesting a point mutation. We have established that this Thr codon is ACG. We can make a Ser codon with a single base pair substitution (UCG). The DNA change would be TA to AT, a transversion.

18.6   In mutant strain X of E. *coli,* a Leu tRNA which recognizes the codon 5'-CUG-3' in normal cells has been altered so that it now recognizes the codon 5'-GUG-3'. A missense mutation, which affects amino acid 10 of a particular protein, is suppressed in mutant X cells.

    a.   What are the anticodons of the two Leu tRNAs, and what mutational event has occurred in mutant X cells?
    b.   What amino acid would normally be present at position 10 of the protein (without the missense mutation)?
    c.   What amino acid would be put in at position 10 if the missense mutation is not suppressed (i.e., in normal cells)?
    d.   What amino acid is inserted at position 10 if the missense mutation is suppressed (i.e., in mutant X cells)?

Answer: a. The normal anticodon was 5'-CAG-3', while the mutant one is 5'-CAC-3'. The mutational event was a CG to GC transversion.

        b. Presumably Leu
        c. Val
        d. Leu

18.7   In any kind of chemotherapy, the object is to find a means to kill the invading pathogen or cancer cell without killing the cells of the host. To do this successfully, one must find and exploit biological differences between target organisms and host cells. Explain the nature of the biological differences between host cells and HIV-1 virus that permits the use of azidothymidine for chemotherapy.

Answer: The reverse transcriptase of HIV-1 recognizes AZT as a substrate (as if it were thymidine), and incorporates it into viral DNA. Cellular DNA polymerases, on the other hand, do not recognize AZT as a substrate, and so it does not become incorporated into cellular DNA.

18.8   The mutant *lac z*-1 was induced by treating *E. coli* cells with acridine, while *lac z*-2 was induced with 5BU. What kinds of mutants are these likely to be? Explain. How could you confirm your predictions by studying the structure of the β-galactosidase in these cells?

Answer: Acridine is an intercalating agent, and so can be expected to induce frameshift mutations. 5BU, on the other hand, is incorporated in place of T, but is relatively likely to be read as C by DNA polymerase because of keto to enol shift. Thus 5BU induced mutations would be expected to be point mutations, usually TA to CG transitions. If these expectations are realized, *lac z*-1 would probably contain a single amino acid difference from the normal β-galactosidase, although it could be a truncated normal protein due to a nonsense point mutation. *lac z*-2, on the other hand, should have a completely altered amino acid sequence after some point, and might also be truncated.

18.9   a. The sequence of nucleotides in an mRNA is:

$$5'\text{-AUGACCCAUUGGUCUCGUUAG-}3'$$

    How many amino acids long would you expect the polypeptide chain made with this messenger to be?
    b. Hydroxylamine is a mutagen that results in the replacement of an AT base pair for a GC base pair in the DNA; that is, it induces a transition mutation. When applied to the organism that made the mRNA molecule shown in part a, a strain was isolated in which a mutation occurred at the 11th position of the DNA that coded for the mRNA. How many amino acids long would you expect the polypeptide made by this mutant to be? Why?

Answer: a. Six.
           b. Three. The mutation results in the replacement of the UGG codon in the mRNA by UAG, which is a chain termination codon.

18.10   In a series of 94,075 babies born in a particular hospital in Copenhagen, 10 were achondroplastic dwarfs (this is an autosomal dominant condition). Two of these 10 had an achondroplastic parent. The other 8 anchondroplastic babies each had two normal parents. What is the apparent mutation rate at the achondroplasia locus?

Answer: There were eight new mutations, out of a total of 2 x 94,075 (or 188,150) copies of the locus, for a mutation rate of 8/188,150 or just over 4×10^{-5} mutations per locus per generation.

18.11   Three of the codons in the genetic code are chain-terminating codons for which no naturally occurring tRNAs exist. Just like any other codons in the DNA, though, these codons can change as a result of base-pair changes in the DNA. Confining yourself to single base-pair changes at a time, determine which amino acids could be inserted in a polypeptide by mutation of these chain-terminating codons: (a) UAG; (b) UAA; (c) UGA. (The genetic code is listed in Figure 14.3.)

Answer: a. UAG: CAG Gln; AAG Lys; GAG Glu; UUG Leu; UCG Ser; UGG Trp; UAU Tyr; UAC Tyr; UAA chain terminating
b. UAA: CAA Gln; AAA Lys; GAA Glu; UUA Leu; UCA Ser; UGA chain terminating; UAU Tyr, UAC Tyr, UAG chain terminating
c. UGA: CGA Arg; AGA Arg; GGA Gly; UUA Leu; UCA Ser; UAA chain terminating; UGU Cys; UGC Cys; UGG Trp

18.12   The amino acid substitutions in the following figure occur in the α and β chains of human hemoglobin. Those amino acids connected by lines are related by single nucleotide changes. Propose the most likely codon or codons for each of the numbered amino acids. (Refer to the genetic code listed in Figure 14.3.)

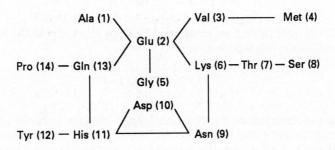

Answer: (1) GCG; (2) GAG; (3) GUG; (4) AUG; (5) GGG; (6) AAG; (7) ACG; (8) UCG; (9) AAU or AAC; (10) GAU or GAC; (11) CAU or CAC; (12) UAU or UAC; (13) CAG; (14) CCG

18.13   Yanofsky studied the tryptophan synthetase of *E. coli* in an attempt to identify the base sequence specifying this protein. The wild type gave a protein with a glycine in position 38. Yanofsky isolated two trp mutants, *A23* and *A46*. Mutant *A23* had Arg instead of Gly at position 38, and mutant *A46* had Glu at position 38. Mutant *A23* was plated on minimal medium, and four spontaneous revertants to prototrophy were obtained. The tryptophan synthetase from each of four revertants was isolated, and the amino acids at position 38 were identified. Revertant 1 had Ile, revertant 2 had Thr, revertant 3 had Ser, and revertant 4 had Gly. In a similar fashion, three revertants from *A46* were recovered, and the tryptophan synthetase from each was isolated and studied. At position 38 revertant 1 had Gly, revertant 2 had Ala, and revertant 3 had Val. A summary of these data is given in the following figure. Using the genetic code shown in Figure 14.3 deduce the codons for the wild type, type, for the mutants *A23* and *A46,* and for the revertants, and place each designation in the space provided in the following figure.

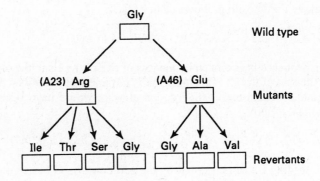

185

Answer:

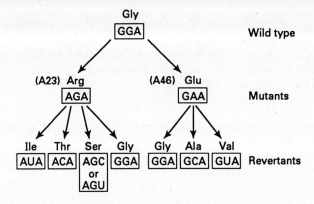

Wild type

(A23) Arg · AGA   (A46) Glu · GAA   Mutants

Ile · AUA   Thr · ACA   Ser · AGC or AGU   Gly · GGA   Gly · GGA   Ala · GCA   Val · GUA   Revertants

18.14   Consider an enzyme chewase from a theoretical microorganism. In the wild-type cell the chewase has the following sequence of amino acids at positions 39 to 47 (reading from the amino end) in the polypeptide chain:

-Met-Phe-Ala-Asn-His-Lys-Ser-Val-Gly-
  39   40   41   42   43   44   45   46   47

A mutant of the organism was obtained; it lacks chewase activity. The mutant was induced by a mutagen known to cause single base-pair insertions or deletions. Instead of making the complete chewase chain, the mutant makes a short polypeptide chain only 45 amino acids long. The first 38 amino acids are in the same sequence as the first 38 of the nominal chewase, but the last 7 amino acids are as follows:

-Met-Leu-Leu-Thr-Ile-Arg-Val-
  39   40   41   42   43   44   45

A partial revertant of the mutant was induced by treating it with the same mutagen. The revertant makes a partly active chewase, which differs from the wild-type enzyme only in the following region:

-Met-Leu-Leu-Thr-Ile-Arg-Gly-Val-Gly-
  39   40   41   42   43   44   45   46   47

Using the genetic code given in Figure 14.3, deduce the nucleotide sequences for the mRNA molecules that specify this region of the protein in each of the three strains.

Answer: For wild-type mRNA:

(39) AUG-UUU-GCU-AAC-CAU-AAG-AGU-GUA-GGX (47)

The second codon (for the amino acid 40) could be UUC. The last codon could be A, G, C, or U at the third position. For mutant mRNA:

(39) AUG-UUG-CUA-ACC-AUA-AGA-GUG-UAG

This sequence is generated by the loss of one of the Us in the second codon (or if it was UUC, by the loss of the C). This mutation is a frameshift mutation and leads to the production of a premature chain-terminating codon at what would have been amino acid 46. For partial revertant mRNA:

(39) AUG-UUG-CUA-ACC-AUA-AGA-GGU-GUA-GGX (47)

The reading frame is restored by the addition of a G before or after the first G in the codon before the chain-terminating codon. This change removes the premature chain-terminating

186

codon and leaves a stretch of six codons that are different from the codons found in the wild-type mRNA.

18.15 Two mechanisms in *E. coli* were described for the repair of DNA damage (thymine dimer formation) after exposure to ultraviolet light: photoreactivation and excision (dark) repair. Compare and contrast these mechanisms, indicating how each achieves repair.

Answer: Photoreactivation needs photolyase enzyme and photon of light in the 320-370-nm wavelength. It causes the direct cleavage of the thymine dimer. Dark repair requires a number of enzymes and does not depend on light. First, an endonuclease makes a single-stranded nick on the 5' side of the dimer; then an exonuclease trims away part of one strand, including the dimer; next, DNA polymerase fills in the single-stranded region in the 5'-to-3' direction. Finally, the gap is sealed by ligase.

18.16 After a culture of *E. coli* cells was treated with 5-bromouracil, it was noted that the frequency of mutants was much higher than normal. Mutant colonies were then isolated and grown and treated with nitrous acid; some of the mutant strains reverted to wild type.
    a.   In terms of the Watson-Crick model, diagram a series of steps by which 5BU may have produced the mutants.
    b.   Assuming the revertants were not caused by suppressor mutations, indicate the steps by which nitrous acid may have produced the back mutations.

Answer: a.   5BU in its normal form is a T analog; in its rare form it resembles C. The mutation is a AT-to-GC transition.

$$\begin{array}{ccccccc} |\ | & & |\ | & & |\ | & & |\ | \\ \text{A-T} & \rightarrow & \text{A-5BU} & \rightarrow & \text{G-5BU} & \rightarrow & \text{G-C} \\ |\ | & & |\ | & & |\ | & & |\ | \end{array}$$

    b.   Nitrous acid can deaminate C to U, so the reversion is to the original AT pair.

18.17 A single, very hypothetical strand of DNA is composed of the base sequence indicated in this figure.

5'–T–HX–U–A–G–BU-enol–2AP–C–BU–X–2AP-imino–3'

In the figure, A indicates adenine, T indicates thymine, G indicates guanine, C denotes cytosine, U denotes uracil, BU is 5-bromouracil, 2AP is 2-aminopurine, BU-enol is a rare tautomer of 5BU, 2AP-imino is a rare tautomer of 2AP, HX is hypoxanthine, and X is xanthine; 5' and 3' are the numbers of the free, OH-containing carbons on the deoxyribose part of the terminal nucleotides.
    a.   Opposite the bases of the hypothetical strand, and using the shorthand of the figure, indicate the sequence of bases on a complementary strand of DNA.
    b.   Indicate the direction of replication of the new strand by drawing an arrow next to the new strand of DNA from part a.
    c.   When postmeitotic germ cells of a higher organism are exposed to a chemical mutagen before fertilization, the resulting offspring expressing an induced mutation are almost always mosaics for wild-type and mutant tissue. Give at least one reason that in the progenies of treated individuals these mosaics are found and not the so-called complete or whole-body mutants.

Answer: a. and b.

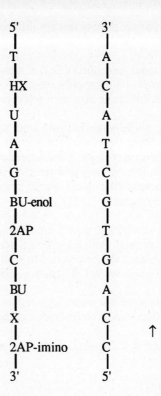

5'	3'
T	A
HX	C
U	A
A	T
G	C
BU-enol	G
2AP	T
C	G
BU	A
X	C
2AP-imino	C
3'	5'

↑

c. One strand is affected by the mutagen. At the first mitotic division one daughter helix consists of two normal strands, and the other daughter helix consists of two mutated strands. Mosaics may be formed when populations of cells arise from these two types by repeated mitoses.

The following information applies to Problems 18.18 through 18.22. A solution of single-stranded DNA is used as the template in a series of reaction mixtures. It has the base sequence as shown in the following figure. The abbreviations are: A = adenine, G = guanine, C = cytosine, T = thymine, H = hypoxanthine, and HNO_2, = nitrous acid. For problems 18.18 through 18.22, use this shorthand system and draw the products expected from the reaction mixtures. Assume that a primer is available in each case.

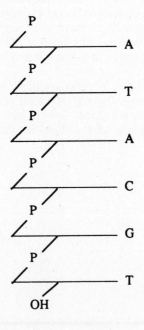

18.18   The DNA template + DNA polymerase + dATP + dGTP + dCTP + dTTP + Mg^{2+}.

Answer:

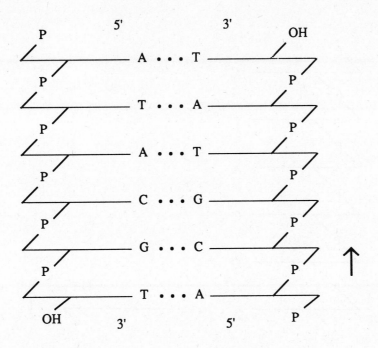

18.19   The DNA template+ DNA polymerase + dATP +dGMP + dCTP + dTTP + Mg^{2+}.

Answer: The reaction stops when it needs dGTP; only dGMP is present.

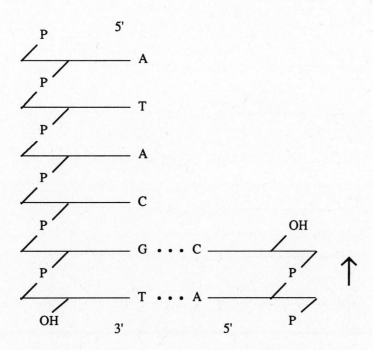

18.20   The DNA template + DNA polymerase + dATP + dHTP + dGMP + dTTP + Mg²⁺.

Answer: dHTP substitutes for the lack of dGTP, but there is no dCTP.

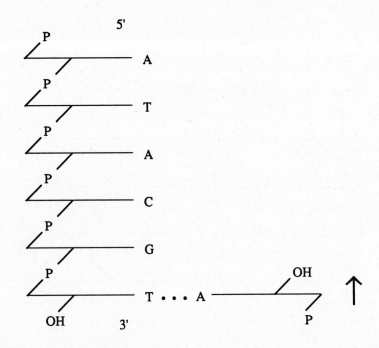

18 21   The DNA template is pretreated with HNO_2 + DNA polymerase + dATP + dGTP + dCTP + dTTP + Mg^{2+}.

Answer: HNO_2 converts A to H, C to U, and G to X.

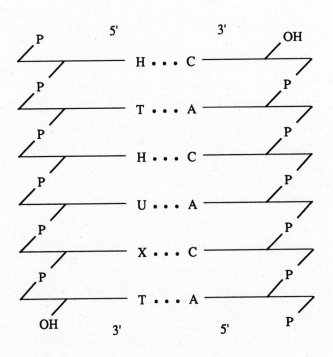

18.22 The DNA template + DNA polymerase+ dATP + dGMP + dHTP + dCTP + dTTP + Mg^{2+}.

Answer: dHTP substitutes for the absence of dGTP.

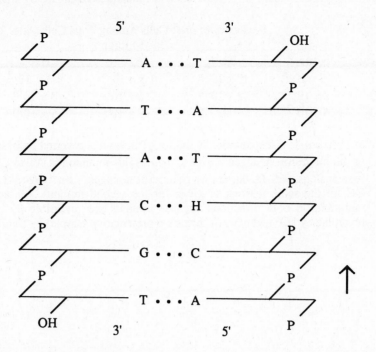

18.23 A strong experimental approach to determining the mode of action of mutagens is to examine the revertibility of the products of one mutagen by other mutagens. The following table represents collected data on revertibility of various mutagens on *rII* mutations in phage T2; + indicates majority of mutants reverted, – indicates virtually no reversion; BU = 5-bromouracil, AP = 2-aminopurine, NA = nitrous acid, and HA = hydroxylamine. Fill in the empty spaces.

Mutation induced by	Proportions of mutations reverted by				Base-pair substitution inferred
	BU	AP	NA	HA	
BU	+	__	__	–	_____
AP	__	–	__	+	_____
NA	+	+	__	+	_____
HA	__	__	+	–	GC → AT

Answer:

Mutation induced by	Proportions of mutations reverted by				Base-pair substitution inferred
	BU	AP	NA	HA	
BU	+	+	+	–	GC → AT
AP	+	–	+	+	AT → GC
NA	+	+	+	+	GC ⇄ AT
HA	–	+	+	–	GC → AT

191

18.24 Three *ara* mutants of *E. coli* were induced by mutagen X. The ability of other mutagens to cause the reverse change (*ara* to *ara*⁺) was tested, with the results shown in the following table.

Assume all ara⁺ cells are true revertants. What base changes were probably involved in forming the three original mutations? What kind(s) of mutations are caused by mutagen X?

Frequency of *ara*⁺ Cells Among Total Cells After Treatment

Mutant	None	BU	AP	HA	Frameshift
ara-1	1.5×10^{-8}	5×10^{-5}	1.3×10^{-4}	1.3×10^{-8}	1.6×10^{-8}
ara-2	2×10^{-7}	2×10^{-4}	6×10^{-5}	3×10^{-5}	1.6×10^{-7}
ara-3	6×10^{-7}	10^{-5}	9×10^{-6}	5×10^{-6}	6.5×10^{-7}

Answer: *ara*⁺ to *ara*-1: This mutation is CG to AT since it is reverted by base analogs but not by HA or the frameshift mutagen. *ara*⁺ to *ara*-2: This mutation is AT to GC since it is reverted by base analogs and HA but not by frameshift mutagen. *ara*⁺ to *ara*-3: This mutation is AT to GC for the same reasons as given for the second mutant. Mutagen X causes transition mutations in both directions because mutants are revertible by base analogs, some are revertible by HA, and none (if this is a representative sample) by frameshift mutagens.

CHAPTER 19

CHROMOSOME ABERRATIONS

I. CHAPTER OUTLINE

Types of Chromosome Changes
Variations in Chromosome Structure
    Deletion
    Duplication
    Inversion
    Translocation
Variations in Chromosome Number
    Changes in One or a Few Chromosomes
    Changes in Complete Sets of Chromosomes
Chromosome Aberrations and Human Tumors
    Chronic Myelogenous Leukemia and the Philadelphia Chromosome
    Chromosomal Aberration and Burkitt's Lymphoma
Chromosome Rearrangements That Alter Gene Expression
    Amplification or Deletion of Genes
    Inversions That Alter Gene Expression
    Transpositions That Alter Gene Expression

II. IMPORTANT TERMS AND CONCEPTS

deletion	aneuploidy
duplication	nullisomy
inversion	monosomy
translocation	trisomy
polytene chromosome	tetrasomy
cri-du-chat syndrome	trisomy 21
reverse tandem duplication	Robertsonian translocation
terminal tandem duplication	polyploidy
unequal crossing-over	triploid
paracentric inversion	tetraploid
pericentric inversion	autopolyploidy
dicentric bridge	allopolyploidy
transposition	gene amplification
Philadelphia chromosome	Burkitt' s Lymphoma

III. THINKING ANALYTICALLY

Perhaps more than any other chapter, this chapter requires the visualization of physical structure, particularly where internal alterations of chromosome structure are concerned, such as inversions, transpositions, and translocations.

Profound changes in gene expression often result from alterations in the physical positions of genes relative to each other or to their control mechanisms, without any other kind of gene change being involved. Disruptions in meiosis in diploid organisms are common sequellae of certain chromosomal aberrations, which are best understood when examined structurally, -as opposed to purely conceptually. This is dramatically illustrated in the case of crossing-over in certain classes of inversions heterozygotes and translocation heterozygotes. One must be prepared to draw diagrams of the chromosomes as a necessary component of one's analytical procedure. Examine carefully problems 19.16 through 19.21 at the end of this chapter in your text, as well as the solutions given in this manual.

IV. QUESTIONS FOR PRACTICE

A. Multiple Choice Questions

1. The number of polytene chromosomes in dipterans, such as *Drosophila*, is characteristically
    a. twice the diploid number.
    b. hundreds or more times the diploid number.
    c. half the diploid number.
    d. the same as the diploid number.

2. The chromosomes that become polytene chromosomes are
    a. somatically paired.
    b. genetically inert.
    c. uniformly pycnotic.
    d. acentric.

3. Nonhomologous chromosomes that have exchanged segments are the products of a
    a. double deletion
    b. reciprocal translocation.
    c. pericentric inversion.
    d. paracentric inversion.

4. The abbreviated karyotype, 2N-1, describes
    a. nullisomy.
    b. monosomy.
    c. trisomy.
    d. haploidy.

5. The abbreviated karyotype 2N+1+1 describes
    a. double trisomy.
    b. tetrasomy.
    c. double monosomy.
    d. none of these.

6. All known monosomics in humans have been
    a. lethal.
    b. semilethal.
    c. treatable.
    d. deletion heterozygotes.

7. The cultivated bread wheat, *Triticum aestivum*, although a polyploid, is fertile because it
    a. has an odd number of chromosome sets.
    b. has an even multiple of chromosome sets.
    c. is a double hybrid.
    d. is propagated by grafting.

8.   Cultivated bananas and Baldwin apples are
      a.   double diploid.
      b.   tetraploid.
      c.   double haploid.
      d.   triploid.
9.   A chromosomal aberration that is the causative agent of chronic myelogenous leukemia is the
      a.   Baltimore chromosome.
      b.   Pittsburgh chromosome.
      c.   Philadelphia chromosome.
      d.   New York chromosome.

10.   Gene expression can be altered by
      a.   an inversion.
      b.   a transposition.
      c.   a duplication.
      d.   any of the foregoing.

Answers: 1c, 2a, 3b, 4b, 5a, 6a, 7b, 8d, 9c, 10d

B.   Thought Questions

1.   Briefly identify and describe some examples of changes in chromosome structure that alter gene expression. (See text, pp. 588-595.)
2.   A chromosomal inversion in the heterozygous condition seems to act as a cross-over suppressor. Explain how this can be. (Hint: what becomes of the cross-over chromatids at anaphase?)
3.   What is both genetically and immunologically noteworthy about the protozoan parasite, *Trypanosoma*, the causative agent of African sleeping sickness? (See text, p. 604.)
4. Describe a Robertsonian translocation and comment on its significance in the inheritance of Down syndrome. (See text, pp. 598-600 and Fig. 19.16.)
5.   How do you account for the fact that polyploidy is significantly more common in plants than in animals? (Hint: what basic difference between sexual and asexual reproduction might be significant here?)
6.   Which do you think would be of greater significance in the evolutionary process of speciation - autopolyploidy or allopolyploidy? Explain. Would your answer be different for animals as compared to plants?

V.   ANSWERS AND SOLUTIONS TO TEXT QUESTIONS

19.1   A normal chromosome has the following gene sequence:

$$A \ B \ C \ D \underset{\circ}{} E \ F \ G \ H$$

Determine the chromosome mutation in each of the following chromosomes:

      a. $A \ B \ C \ F \ E \underset{\circ}{} D \ G \ H$
      b. $A \ D \underset{\circ}{} E \ F \ B \ C \ G \ H$
      c. $A \ B \ C \ D \underset{\circ}{} E \ F \ E \ F \ G \ H$
      d. $A \ B \ C \ D \underset{\circ}{} E \ F \ F \ E \ G \ H$
      e. $A \ B \ D \underset{\circ}{} E \ F \ G \ H$

   a. pericentric inversion [D (centromere) E F inverted]
   b. nonreciprocal translocation (BC moved to other arm)
   c. tandem duplication (EF duplicated)
   d. reverse tandem duplication (EF duplicated)
   e. deletion (C deleted)

195

19.2   Define pericentric and paracentric inversions.

Answer: See text, pp. 592-593.

19.3   Very small deletions behave in some instances like recessive mutations. Why are some recessive mutations known not to be deletions?

Answer: Deletions do not revert. So if a recessive mutation reverts, it cannot be the result of a deletion.

19.4   What would be the result, in terms of protein structure, if a small inversion were to occur within the coding region of a structural gene?

Answer: This would be a serious problem. Because of the 5'- 3' polarity of the two strands, the bases of the anti-sense strand would be inserted into the sense strand in the inverted region. This would result in a completely deranged amino acid sequence, as well as possible nonsense codons.

19.5   Inversions are said to affect crossing-over. The following homologs with the indicated gene order are given:

    A B C D E

    A D C B E

a.   Diagram the way these chromosomes would align in meiosis.
b.   Diagram what a single crossover between homologous genes *B* and *C* in the inversion would result in.
c.   Considering the position of the centromere, what is this sort of inversion called?

Answer:

b.   From part a the crossover between B and C is as follows:

    A B C D E

    A B C D A

    E B C D E

    A D C B E

c.   Paracentric inversion, because the centromere is not included in the inverted DNA segment.

196

19.6 Single crossovers within the inversion loop of inversion heterozygotes give rise to chromatids with duplications and deletions. What happens when, within the inversion loop, there is a twostrand double crossover in such an inversion heterozygote when the centromere is outside the inversion loop?

Answer: An example of a two-strand double crossover and the resulting meiotic products is shown in the following figure:

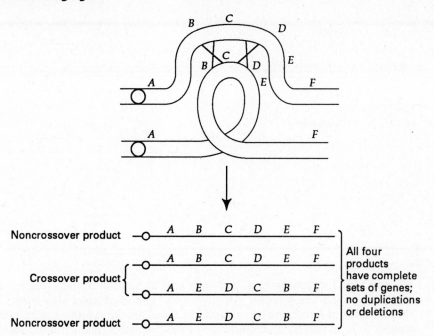

In this case no dicentric bridge structure is formed, as is the case with single crossovers, and all four meiotic products are viable.

19.7 An inversion heterozygote possesses one chromosome with genes in the normal order:

$$a\ b\ c\ d\ e\ f\ g\ h$$

It also contains one chromosome with genes in the inverted order:

$$a\ b\ f\ e\ d\ c\ g\ h$$

A four-strand double crossover occurs in the areas *e-f* and *c-d*. Diagram and label the four strands at synapsis (showing the crossovers) and at the first meiotic anaphase.

Answer: In the following, the crossover between *c* and *d* involves strands 2 and 4, and the crossover between *e* and *f* involves strands 1 and 3.

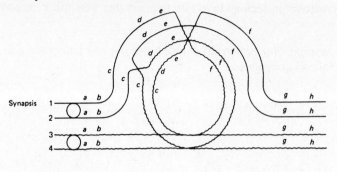

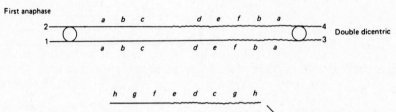

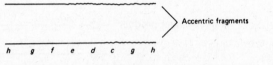

19.8 Mr. and Mrs. Lambert have not yet been able to produce a viable child. They have had two miscarriages and one severely defective child who died soon after birth. Studies of banded chromosomes of father, mother, and child showed all chromosomes were normal except for pair number 6. The number 6 chromosome of mother, father, and child are shown in the following figure.

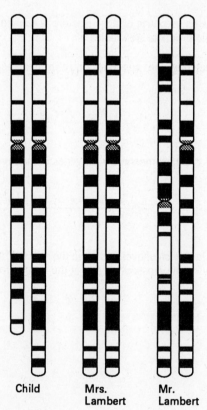

a. Does either parent have an abnormal chromosome? If so, what is the abnormality?
b. How did the chromosomes of the child arise? Be specific as to what events in the parents gave rise to these chromosomes.
c. Why is the child not phenotypically normal?
d. What can be predicted about future conceptions by this couple?

Answer: a. Mr. Lambert is heterozygous for a pericentric inversion of chromosome 6. One of the breakpoints is within the fourth light band up from the centromere, and the other is in the sixth dark band below the centromere. Mrs. Lambert's chromosomes are normal.

b. When Mr. Lambert's chromosomes 6s paired, they formed an inversion loop which included the centromere. Crossing over occurred within the loop, and gave rise to the partially duplicated, partially deficient 6 which the child received.

c. The child's abnormalities stem from having three copies of some, and only one copy of other, chromosome 6 regions. The top part of the short arm is duplicated, and there is a deficiency of the distal part of the long arm in this case.

d. The inversion appears to cover more than. half of the length of chromosome 6, so crossing over will occur in this region in the majority of meioses. In the minority of meioses where crossing over occurs outside the loop, and in the cases where it has occurred within but the child receives an uncrossed over chromatid, the child can be normal. There is significant risk for abnormality, so monitoring of fetal chromosomes should be done.

19.9   Mr. and Mrs. Simpson have been trying for years to have a child but have been unable to conceive. They consulted a physician, and tests revealed that Mr. Simpson had a very reduced sperm count. His chromosomes were studied, and a testicular biopsy was done as well. His chromosomes proved to be normal, except for pair 12. The following figure shows a normal pair of number 12 chromosomes and Mr. Simpson's number 12 chromosomes.
a. What is the nature of the abnormality in Mr. Simpsons chromosome 12s?
b. What abnormal feature would you expect to see in the testicular biopsy (cells in various stages of meiosis can be seen).
c. Why is Mr. Simpson's sperm count low?
d. What can be done about it?

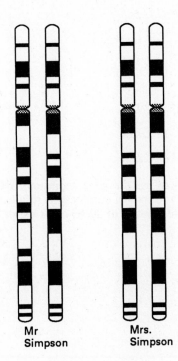

Mr Simpson   Mrs. Simpson

Answer: a. He has pararacentic inversion within the long arm of one of his number 12 chromosomes.
b. Crossing over within the inversion loop should produce dicentric chromatids which would form anaphase bridges. These chromatin bridges would be visible joining the two chromatin masses at anaphase I.
c. The inversion is large, so bridges will be formed in the majority of meioses. Cells which form a bridge do not complete meiosis or form sperm in mammals.
d. Nothing can be done to increase his sperm count.

19.10 The following gene arrangements in a particular chromosome are found in *Drosophila* populations in different geographical regions. Assuming the arrangement in part a is the original arrangement, in what sequence did the various inversion types most likely arise?
a. *ABCDEFGHI*
b. *HEFBAGCDI*
c. *ABFEDCGHI*
d   *ABFCGHEDI*
e. *ABFEHGCDI*

Answer:

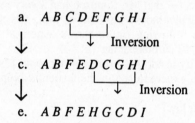

That is, the following sequence occurred:

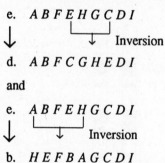

d and b are derived from e by separate inversion events:

e. *A B F E H G C D I*
↓   ↓ Inversion
d. *A B F C G H E D I*

and

e. *A B F E H G C D I*
↓   ↓ Inversion
b. *H E F B A G C D I*

19.11 Chromosome I in maize has the gene sequence *ABCDEF*, whereas chromosome II has the sequence *MNOPQR*. A reciprocal translocation resulted in *ABCPQR* and *MNODEF*. Diagram the expected pachytene configuration of the F_1 of a cross of homozygotes of these two arrangements.

200

Answer:

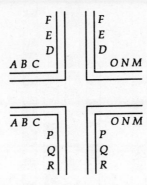

19.12. Diagram the pairing behavior at prophase of meiosis I of a translocation heterozygote with normal chromosomes of gene order *abcdefg and tuvwxyz* and the translocated chromosomes *abcdvwxyz* and *tuefg*. Assume that the centromere is at the left end of all chromosomes.

Answer:

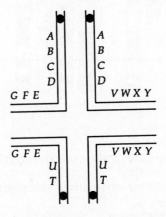

19.13 Mr. and Mrs. Denton have been trying for several years to have a child. They have experienced a series of miscarriages, and last year they had a child with multiple congenital defects. The child died within days of birth. The birth of this child prompted the Denton's physician to order a chromosome study of parents and child. The results of the study are shown in the following figure. Chromosome banding was done, and all chromosomes were normal in these individuals except some copies of number 6 and number 12. The number 6 and number 12 chromosomes of mother, father, and child are shown in the figure (the number 6 chromosomes are the larger pair).

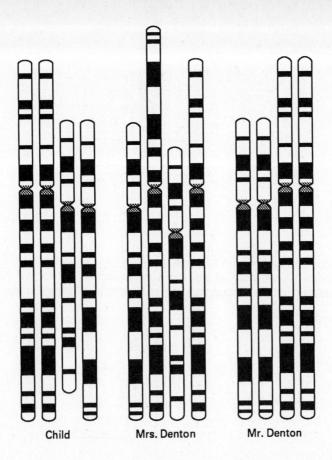

Child   Mrs. Denton   Mr. Denton

a. Does either parent have an abnormal karyotype? If so, which parent has it, and what is the nature of the abnormality?

b. How did the child's karyotype arise (what pairing and segregation events took place in the parents)?

c. Why is the child phenotypically defective?

d. What can this couple expect to occur in subsequent conceptions?

e. What medical help, if any, can be offered to them?

Answer: a. Mr. Denton has normal chromosomes. Mrs. Denton is heterozygous for a balanced reciprocal translocation between chromosomes 6 and 12. Most of the short arm of chromosome 6 has been reciprocally translocated onto the long arm of chromosome 12. The breakpoints appear to be in the thick, dark band just above the centromere of 6 and in the third dark band below the centromere of 12.

         b. The child received a normal 6 and a normal 12 from his father. In meiosis in Mrs. Denton, the 6s and 12s paired and formed a cross-like figure in prophase I. Segregation of adjacent nonhomologous centromeres to the same pole ensued, and the child received a gamete containing a normal 6 and one of the translocation chromosomes (see Figure 19.12).

         c. The child received a normal chromosome 6 and a normal 12 from Mr. Denton. The child is not physically normal because of partial trisomy (it has three copies of part of the short arm of chromosome 6), and partial monosomy (it has only one copy of most of the long arm of chromosome 12).

         d. Given that segregation of adjacent homologous centromeres to the same pole will be relatively rare in this case, there should be about a 50% chance that a given conception will be chromosomally unbalanced. Of the 50% balanced ones, 50% would be translocation heterozygotes.

         e. Prenatal monitoring of fetal chromosomes could be done, followed by therapeutic abortion of chromosomally unbalanced fetuses.

19.14   Define the terms *aneuploidy*, *monoploidy*, and *polyploidy*.

Answer: See text, pp. 597, 601, 602.

19 15   If a normal diploid cell is 2N, what is the chromosome content of the following: (a) a nullisomic; (b) a monosomic; (c) a double monosomic; (d) a tetrasomic; (e) a double trisomic; (f) a tetraploid; (g) a hexaploid?

Answer: a. 2N - 2
              b. 2N - 1
              c. 2N - 1 - 1
              d. 2N + 2
              e. 2N + 1 + 1
              f. 4N
              g. 6N

19.16   In humans how many chromosomes would be typical of nuclei of cells that are (a) monosomic; (b) trisomic; (c) monoploid; (d) triploid; (e) tetrassomic?

Answer: a. 45
              b. 47
              c. 23
              d. 69
              e. 48

19.17   An individual with 47 chromosomes, including an additional chromosome 15, is said to be:
        a.   A triplet.
        b.   Trisomic.
        c.   Triploid.
        d.   Tricycle.

Answer: b. An individual with three instead of two chromosomes is said to be *trisomic*.

19.18   A color-blind man marries a homozygous normal woman and after four joyful years of marriage they have two children. Unfortunately, both children have Turner -syndrome, although one has normal vision and one is color blind. The type of color-blindness involved is a sex-linked recessive trait.
        a.   For the color-blind child with Turner syndrome: did non-disjunction occur in the mother or the father? Explain your answer.
        b.   For the Turner child with normal vision in which parent did non-disjunction occur.? Explain your answer.

Answer: a. Mother. The color-blind Turner syndrome child must have received her X chromosome from her father since the father carries the only color-blind allele. Therefore, the mother must have produced an egg with no X chromosomes.
              b. Father. Since the Turner syndrome child does not have color-blindness, she must have received her X chromosome from the homozygous normal mother. Therefore, the father must have produced a sperm with no X or Y chromosome.

19.19   In general, polyploids with even multiples of the chromosome set are more fertile than polyploids with odd multiples of the chromosome set. Why?

Answer: Polyploids with even multiples of the chromosomes can better form chromosome pairs in meiosis than can polyploids with odd multiples. Triploids, for example, will generate an unpaired chromatid pair for each chromosome type in the genome, so chromosome segregation to the gametes is irregular.

19.20　From Mendel's first law genes *A* and *a* segregate from each other and appear in equal numbers among the gametes. However obvious that phenomenon may seem now, it was not obvious to Mendel. Mendel did not know that his plants were diploid. In fact, since plants are frequently tetraploid, he could have been unlucky enough to have started with peas that were 4N rather than 2N. Let us assume that Mendel's peas were tetraploid, that every gamete contains two alleles, and that the distribution of alleles to the gamete is random. Suppose we have a cross of *AA AA* x *aa aa*, where *A* is dominant regardless of the number of *a* alleles present in an individual.
　　　a.　What will be the genotype of the F_1?
　　　b.　If the F_1 is selfed, what will be the phenotypic ratios in the F_2?

Answer:　　a.　*AA aa*
　　　　　　b.　If we label the four alleles *A1*, *A2*, *a1*, and *a2*, there are 6 possible gamete types: *A1 A2*, *A1 a1*, *A1 a2*, *A2 a1*, *A2 a2*, and *a1 a2*, i.e., 1/6 *AA*, 4/6 *Aa* and 1/6 *aa*. The possible gamete pairings are as follows:

	1/6 *AA*	4/6 *Aa*	1/6 *aa*
1/6 *AA*	1/36 *AAAA*	4/36 *AAAa*	1/36 *AAaa*
4/6 *Aa*	4/36 *AAAa*	16/36 *AAaaa*	4/36 *Aaaa*
1/6 *aa*	1/36 *AAaa*	4/36 *Aaaa*	1/36 *aaaa*

　　　　　　Phenotypically this gives 35/36 *A* : 1/36 *a*.

19.21　What phenotypic ratio of *A* to *a* is expected if *AA aa* plants are testcrossed against *aa aa* individuals? (Assume that the dominant phenotype is expressed whenever at least one *A* is present, no crossing-over occurs, and each gamete receives two chromosomes.)

Answer:　　5 *A*:1 *a*

19.22　The root tip cells of an autotetraploid plant contain 48 chromosomes. How many chromosomes did the gametes of the diploid from which this plant was derived contain?

Answer:　　12

19.23　How many chromosomes would be found in somatic cells of an allotetraploid derived from two plants, one with N = 7 and the other with N = 10?

Answer:　　(7N + 10N) x 2 = 34 (4N)

19.24　Plant species A has a haploid complement of 4 chromosomes. A related species B has N = 5. In a geographical region where A and B are both present, C plants are found that have some characters of both species and somatic cells with 18 chromosomes. What is the chromosome constitution of the C plants likely to be? With what plants would they have to be crossed in order to produce fertile seed?

Answer:　　The C Plants are allotetraploids, containing a diploid set of chromosomes from both A and B. Fertile seed will be produced when they are crossed with other C plants because no abnormnal chromosome pairing or unpaired chromosomes will result in that case.

CHAPTER 20

TRANSPOSABLE GENETIC ELEMENTS

I. CHAPTER OUTLINE

Transposable Genetic Elements in Bacteria
    Insertion Sequence
    Transposons
    IS Elements and Transposons in Plasmids
    Temperate Bacteriophage *Mu*
Transposable Elements in Eukaryotes
    Transposons in Plants
    Transposons in Yeast
    Drosophila transposons
Tumor Viruses, Oncogenes, and Cancer
    Tumor Viruses
    Retroviruses
    HIV – The AIDS Virus

II. IMPORTANT TERMS AND CONCEPTS

oncogene *vs.* proto-oncogene	*Ty* element
IS element *vs. IS* module	LTRs and deltas
transposon	retrovirus *vs.* retrotransposon
transposition event	provirus
transposase *vs.* resolvase	reverse transcriptase
stem and loop structure	*copia* retrotransposon
direct *vs.* inverted repeat	tumor *vs.* cancer
conservative *vs.* replicative	metastasis
transposition	oncogenesis
plasmid *vs.* episome	*src*
R plasmid group	RSV
RTF region	v-*onc vs.* c-*onc*
null mutation	transducing retrovirus
autonomous *vs.* nonautomous element	HIV
stable *vs.* unstable mutant	AIDS
Ac-Ds family	protein kinase

III. THINKING ANALYTICALLY

Since the material in this chapter is broadly descriptive, it is important to distinguish between the characteristics of the various categories of transposable genetic elements, and to understand the differences and associations between viral and cellular phenomena.

    The distinction between direct and inverted repeats is central to the recognition of insertion sequences. The insertion sequence itself contains inverted terminal repeats. Thus, inverted repeats are present as a part of the transposable element both before and after insertion. On the other hand, the act of insertion results in a staggered cut in the host DNA. When the gaps are filled in after insertion, direct repeats will be generated flanking the inverted terminals of the transposable element. Indeed, insertion sequences have been identified by the presence of a pair of inverted repeats flanked by direct repeats. Using this logic, find the insertion sequence in the following single-stranded length of DNA:

CGTAGCCATTTGCGATATGCATCCGAATATCGCAAGCCATGCCA

Answer: CGTAGCCATTTGCGATATGCATCCGAATATCGCAAGCCATGCCA

Insertion
sequence

IV. QUESTIONS FOR PRACTICE

A. Multiple Choice Questions

1. Transposition and recombination are fundamentally different activities because:
    a. Recombination requires DNA homology but transposition does not.
    b. Transposition requires DNA homology but recombination does not.
    c. Only during transposition can a piece of DNA be moved from a virus to a cell.
    d. Only during recombination can a deletion error occur.

2. Which of the following can result from the insertion of a transposon in bacteria?
    a. Gene inactivation.
    b. Depression or enhancement of gene activity.
    c. Deletions and insertions.
    d. All of the preceding.

3. Transposition has been known to cause which of the following in eukaryotes?
    a. Conversion of a heterozygote from the dominant to the recessive phenotype.
    b. Production of a null mutation.
    c. Increased or decreased efficiency of promoter regions.
    d. All of the preceding.

4. What are the causes and conditions that result in the formation of a "lollipop" (stem and loop) structure of an prokaryotic insertion sequence?
    a. Inverted terminal repeats.
    b. Direct terminal repeats.
    c. Denaturation to separate the strands of DNA.
    d. More than one of the preceding.

5. Transposition in bacteria is a rare event because:
    a. There are very few transposons found in bacteria.
    b. Transposase, the enzyme required for transposition, is made very slowly.
    c. The origination of the pattern of direct and inverse repeats can occur only infrequently.
    d. The ligases that are required are in short supply.

6. Why does the insertion of a autonomous element result in the production of an unstable (mutable) allele?
    a. Although an autonomous element can not excise, it becomes a mutable hot-spot.
    b. An autonomous element can excise, thereby causing a mutation.
    c. Autonomous alleles will be unstable when in the company of a non-autonomous element.
    d. Autonomous alleles generally insert into promoter regions.

7. *Ty* elements are more similar to retroviruses than to transposons in corn because:
   a. Both *Ty* elements and retroviruses utilize an RNA intermediate.
   b. Although all three kinds of transposons synthesize a reverse transcriptase, that enzyme in corn never becomes functional.
   c. Introns are usually found in retroviruses as well as *Ty* elements, but they are not found in plant transposons.
   d. More than one of the preceding.

8. Which of the following transposons transpose via an RNA intermediate?
   a. *IS* sequences and plasmids.
   b. Temperate bacteriophage and plasmids.
   c. *Ty, copia* elements, and retroviruses.
   d. All of the preceding.

9. From the list of characteristics below, select the one which is minimally necessary for a retrovirus to be a tumor-inducing virus.
   a. It must contain the v-*onc* gene.
   b. It must code for reverse transcriptase.
   c. Helper viruses must be present.
   d. Proto-oncogenes must be present.

10. Which of the following characteristics are present in cellular proto-oncogenes, but not viral oncogenes?
    a. Cellular proto-oncogenes may contain introns.
    b. Cellular proto-oncogenes are less able to replicate.
    c. Proto-oncogenes generate larger amounts of mRNA.
    d. All of the preceding are found in both.

Answer: 1a, 2d, 3d, 4d, 5b, 6b, 7a, 8c, 9a, 10a

B. Thought Questions

1. Contrast conservative with replicative transposition, using the *Ac* element in corn to show how conservative transposition can result in gene duplication. Can particular *IS* sequences behave in both ways? (See text, pp. 614, 621-624.)
2. Compare the insertion of antibiotic resistance factors in *Shigella* with that of the fertility factor in *E. coli*. (See text pp. 614-619.)
3. How can transposable genetic elements create null mutations on the one hand, as well as mutations involving increased gene function on the other? (See text, pp. 623.)
4. When can a non-autonomous element behave as an autonomous element? (See text, p. 623)
5. Contrast those transposable genetic elements that utilize an RNA intermediate with those that do not.

V. ANSWERS AND SOLUTIONS TO TEXT QUESTIONS

20.1   Compare and contrast the four types of transposable genetic elements in prokaryotes.

Answer: The four types are insertion sequences, transposons, plasmids, and certain temperate bacteriophages. See text pp. 612-620 for discussion.

20.2   What are the properties in common between prokaryotic and eukaryotic transposons?

Answer: See text, pp. 612-620.

20.3   An *IS* element became inserted into the *lacZ* gene of *E. coli*. Later, a small deletion occurred in this gene, which removed forty base pairs, starting to the left of the *IS* element. Ten *lacZ* base pairs were removed, including the left copy of the target site, and the thirty left most base pairs of the *IS* element were removed. What will be the consequence of this deletion?

Answer: The two ends of this *IS* element are no longer homologous. It will not be able to move out of this location and insert at another site.

20.4   A molecular biologist was studying the DNA of a newly discovered fungus, using denaturation-renaturation experiments to gain information about the amount of single-copy *vs.* repeated sequences. He was surprised to discover that, while most of the DNA of this fungus fit the usual $C_o t$ expectations (renaturation time inversely related to concentration), 2% of the DNA failed to behave this way. This 2% always renatured at the same rate (very quickly) irrespective of concentration (even when the DNA solution was very dilute). Suggest an explanation for this observation.

Answer: Renaturation is ordinarily dependent on concentration because two single stranded molecules must collide in order for renaturation to begin. Obviously the probability of collision increases as concentration increases. If renaturation of a particular kind of DNA is independent of concentration, it means that collision is not required for renaturation to occur. This in turn means that the renaturing sequences are on the same single stranded molecule, indicating closely spaced inverted repeats. Thus the results indicate that 2% of the DNA of this fungus contains closely spaced inverted repeats, some of which may represent transposable elements.

20.5   A geneticist was studying glucose metabolism in yeast, and had deduced both the normal structure of the enzyme glucose-6-phosphatase (G6Pase) and the DNA sequence of its coding region. She had been using a wild-type strain called A to study another enzyme for many generations, when she notice a morphologically peculiar mutant had arisen from one of the strain A cultures. She grew the mutant up into a large stock and found that the defect in this mutant involved a markedly reduced G6Pase activity. She isolated the G6Pase protein from these mutant cells and found it was present in normal amounts, but had an abnormal structure. The N-terminal 70% of the protein was normal. The C-terminal 30% was present but altered in sequence by a frame shift reflecting the insertion of 1 base pair, and the N-terminal 70% and the C-terminal 30% were separated by 111 new amino acids unrelated to normal G6Pase. These amino acids represented predominantly the AT rich codons (Phe, Leu, Asn, Lys, Ile, Tyr). There were also two extra amino acids added at the C-terminal end. Explain these results.

Answer: The extra 111 amino acids plus the one base frame shift indicate the insertion of 334 base pairs into the G6Pase structural gene. This is consistent with the idea that a *Ty* element was inserted into the G6Pase gene, and then one of its deltas recombined with another delta, excising the *Ty* element and leaving delta behind in the G6Pase gene. Delta is 334 base pairs. If this were to be positioned so that it would be translated and not generate a nonsense codon, it would yield 111 amino adds and one extra base pair, which would cause the frame shift. The extra amino acids at the end of the molecule presumably occurred because the frame shift did not allow the normal termination codon to be read.

20.6   Consider two theoretical yeast transposons, A and B. Each contains an intron. Each transposes to a new location in the yeast genome and then is examined for the presence of the intron. In the new locations, you find that A has no intron, while B does. What can you conclude about the mechanisms of transposon movement for A and B from these facts?

Answer: Since a transposition results in the loss of the intron, it may be hypothesized that transposition of A occurs via an RNA intermediate. That is, intron removal occurs only at the RNA level.

        The lack of intron removal during B transposition suggests that there is a DNA - DNA transposition mechanism, without any RNA intermediate.

20.7   An investigator has found a retrovirus capable of infecting human nerve cells. This is a complete virus, capable of reproducing itself, and it contains no oncogenes. People who are infected suffer a debilitating encephalitis. The investigator has shown that when the infects nerve cell in culture with the complete virus, the nerve cells and killed as the virus reproduces, but if he infects cultured nerve cells with a virus in which he has created deletions in the *env* or *gag* genes, no cell death occurs. The investigator is interested in finding ways to bring about nerve cell growth or regeneration in people who have suffered nerve damage. For example, in a patient with a severed spinal cord, nerve regeneration might relieve paralysis. The investigator has cloned the human nerve growth factor gene, and wants to insert it into the genome of his retrovirus from which he has deleted parts of the *env* and *gag* genes. He would then use the engineered retrovirus to infect-cultured nerve cells. Adult nerve cells do not normally produce large amounts of nerve growth factor. If he is successful in inducing growth in them without causing any cell death, he would like to move on to clinical trials on injured patients. When the investigator applied for grant support to do this work, his application was denied on grounds that there were inadequate safeguards in the plan. Why might this work be dangerous? What comparisons can you \draw between the virus the investigator wants to create and, for example, Avian myeloblastosis virus?

Answer: In engineering the retrovirus in the way he plans, the investigator would probably be creating a new cancer virus, in which the cloned nerve growth factor gene would be the oncogene. It is, of course, an advantage that the engineered virus would not be able to reproduce itself, but we know that many "wild" cancer viruses are also defective and reproduce with the help of other viruses. If the engineered virus were to infect cells carrying other viruses (for example wild-type versions of itself) which could supply the *env* and *gag* functions, the new virus could be reproduced and spread. Presumably, infection of normal nerve cells *in vivo* by the engineered, retrovirus would sometimes result in abnormally high levels of nerve growth factor, and thus perhaps to the production of nervous system cancers.

In Avian myeloblastosis virus (AMV) the *pol* and *env* genes are partially deleted. Thus, like our investigator's virus, AMV needs a helper virus to reproduce. In AMV the *myb* oncogene has been inserted, which encodes a nuclear protein presumably involved in control of gene expression. In our new virus the oncogene would be the cloned nerve growth factor gene.

CHAPTER 21

ORGANIZATION AND GENETICS OF EXTRANUCLEAR GENOMES

I. CHAPTER OUTLINE

Organization of Extranuclear Genomes
    Mitochondrial Genomes
    Chloroplast Genomes
Rules of Extranuclear Inheritance
Maternal Effect
Examples of Extranuclear Inheritance
    Leaf Variegation in the Higher Plant *Mirabilis jalapa*
    The *poky* Mutant of *Neurospora*
    Yeast *petite* Mutants
    Extranuclear Genetics of *Chlamydomonas*
    Infectious Heredity – Killer Yeast

II. IMPORTANT TERMS AND CONCEPTS

non-Mendelian inheritance	maternal inheritance
extranuclear inheritance	maternal effect
uniparental inheritance	infectious inheritance
biparental inheritance	killer yeasts

III. THINKING ANALYTICALLY

Review Q1 and Q2 in the Analytical Approaches section at the end of this chapter.

IV. QUESTIONS FOR PRACTICE

A. Multiple Choice Questions

1. DNA is found in
   a. the nucleus of eukaryotes.
   b. mitochondria.
   c. chloroplasts.
   d. all the foregoing.

2. Extranuclear genes in general are characterized by
   a. maternal inheritance.
   b. Mendelian inheritance.
   c. reverse transcription.
   d. the maternal effect.

3. Mitochondrial DNA is
   a. typically single-stranded and coiled in nucleosomes.
   b. uniform in size in all species in which they occur.
   c. uniformly double-stranded, supercoiled and circular.
   d. found only in animals and fungi.

4. Mitochondrial DNA is usually
   a. contained in nucleoid regions.
   b. designated as extranuclear DNA.
   c. replicated synchronously with nuclear DNA.
   d. replicated conservatively.

5. Most of the ribosomal proteins of mitochondrial ribosomes are coded for in the
   a. nucleoid region.
   b. nuclear genome.
   c. H (heavy) strand.
   d. L (light) strand.

6. The mitochondrial mRNAs have
   a. no poly(A) tail.
   b. a 5'cap.
   c. a poly(A) tail but no 5' cap.
   d. both a poly(A) tail and a 5' cap.

7. Mitochondrial DNA uses
   a. the same genetic code as nuclear DNA.
   b. a somewhat different code than nuclear DNA.
   c. a very wobbly code for protein synthesis.
   d. none of the foregoing.

8. The replication of cpDNA is
   a. semiconservative.
   b. dispersive.
   c. conservative.
   d. all the foregoing.

9. The process of protein synthesis in chloroplasts is similar to that in
   a. eukaryotes.
   b. mitochondria.
   c. prokaryotes.
   d. fungi.

10. In some cases, extranuclear inheritance may be due to
   a. symbiotic bacteria.
   b. symbiotic viruses.
   c. either a or b.
   d. neither a nor b

Answers: 1d, 2a, 3c, 4a, 5b, 6c, 7b, 8a, 9c, 10c

B. Thought Questions

1. One of the tenets of the symbiosis theory of the origin of eukaryotic cells is that mitochondria were at one time free-living prokaryotic cells that secondarily became included within the the cytoplasm of other cells, possibly nucleated, where they established a symbiotic relationship that persists to this day in the cells of all eukaryotic organisms. Discuss whatever evidence you can muster, both pro and con, relative to this theory.

211

2. In what ways does cpDNA differ from nuclear and mitochondrial DNA? (See text, pp. 644-650, 651, 653, Table 21.1)

3. How do you account for the maternal effect in shell coiling in the snail, *Limnaea peregra* ?

4. Describe and discuss infectious heredity, citing examples. (See text, pp. 665-666.)

5. Speculate on the evolutionary origin of plant chloroplasts, compared to the presumed origin of mitochondria, described above.

V. ANSWERS AND SOLUTIONS TO TEXT QUESTIONS

21.1   Compare and contrast the structure of the nuclear genome, the mitochondrial genome, and the chloroplast genome.

Answer: The nuclear genome is organized into linear structures called chromosomes, which are composed of double-stranded DNA complexed with histones and nonhistone proteins. The nuclear chromosomes replicate at a specific time in the cell cycle and are segregated to progeny cells in an orderly fashion during the processes of mitosis and meiosis. The mitochondrial and chloroplast genomes consist of circular, naked, double-stranded DNA that is replicated independently of the nuclear genome and in many phases of the cell cycle. Segregation of progeny double helices to daughter organelles does not involve either mitotic or meiotic processes.

21 2   How do mitochondria reproduce? What is the evidence for the method you describe?

Answer: New mitochondria are generated from old mitochondria. See text, p. 644.

21.3   What genes are present in the human mitochondrial genome?

Answer: See text, pp. 644-651.

21.4   What conclusions can you draw from the fact that most nuclear-encoded mRNAs and all mitochondrial mRNAs have a poly(A) tail at the 3' end?

Answer: The exact function of a poly(A) tail on an mRNA molecule remains a mystery. No mitochondrial mRNAs are known to exit the mitochondrion, so we can conclude that the poly(A) tail may not be needed for transport between cellular compartments. Presumably, some basic function is provided by the poly(A) tail. This hypothetical function apparently is not needed by histone messengers.

21 5   Discuss the differences between the universal genetic code of the nuclear genes and the code found in mammalian and fungal mitochondria. Is there any advantage to the mitochondrial code?

Answer: The two codes are different in codon designations. The mitochondrial code has more extensive wobble so that fewer tRNAs are needed to read all possible sense codons. As a consequence, fewer mitochondrial genes are necessary. The advantage is that fewer tRNAs are needed, and hence fewer tRNA genes need be present than is the case for cytoplasmic tRNAs.

21.6   When the DNA sequences for most of the mRNAs in human mitochondria are examined, no nonsense codons are found at their terminal. Instead, either U or UA is found. Explain this result.

Answer: All mitochondrial mRNAs have a poly(A) tail added at their 3' ends after transcription. The string of A's added to the ending U or UA of the transcript completes the chain-terminating codon UAA.

21.7   Compare and contrast the cytoplasmic and mitochondrial protein-synthesizing systems.

Answer: See text, pp. 415-423; 651.

21.8 Compare and contrast the organization of the ribosomal RNA genes in mitochondria and in chloroplasts.

Answer: See text, pp. 647-651; 651-654.

21.9 What features of extranuclear inheritance distinguish it from the inheritance of nuclear genes?

Answer: The features of extranuclear inheritance are differences in reciprocal-cross results (not related to sex), non-mappability to known nuclear chromosomes, Mendelian segregation not followed, and indifference to nuclear substitution.

21.10 Distinguish between maternal effect and extranuclear inheritance.

Answer: Maternal effect (also called predetermination) is the determination of gene-controlled characters by the maternal genotype prior to the fertilization of the egg cell. The genes involved are nuclear genes, whereas the genes involved in extranuclear inheritance are extranuclear, being found in the mitochondria and chloroplasts.

21.11 Distinguish between nuclear (segregational), neutral, and suppressive *petite* mutants of yeast.

Answer: Nuclear *petites* owe their phenotype to nuclear gene mutations, whereas neutral and suppressive *petites* owe their phenotypes to mutational changes in the mitochondrial genome.

21.12 The inheritance of the directrion of shell coiling in the snail *Limneae peregra* has been studied extensively. A snail produced by a cross between two individuals has a shell with a right-hand twist (dextral-coiling). This snail produces only left-hand (sinistral) progeny on selfing. What are the genotypes of the F_1 snail and its parents?

Answer: The parental snails were D/d female and $d/-$ male. The F_1 snail is d/d. (Given the F_1 genotype, the male can either be homozygous or heterozygous, but the determination cannot be made from the data given.)

21.13 *Drosophila melanogaster* has a sex-linked, recessive, mutant gene called *maroon-like (ma-l)*. Homozygous *ma-l* females or hemizygous *ma-l* males have light-colored eyes owing to the absence of the active enzyme xanthine dehydrogenase, which is involved in the synthesis of eye pigments. When heterozygous $ma-l^+/ma-l$ females are crossed with *ma-l* males, all the offspring are phenotypically wild-type. However, half the female offspring from this cross, when crossed back to *ma-l* males, give all *ma-l* progeny. The other half of the females, when crossed to *ma-l* males, give all phenotypically wild-type progeny. What is the explanation for these results?

Answer: There is a maternal effect In the $ma-l^+/ma-l$ female heterozygote, the xanthine dehydrogenase mRNA made by the wild-type allele is pumped into all the eggs, so all the offspring are phenotypically wild-type. (The mRNA must be stable.) Genotypically, the progeny from the cross are half homozygous *ma-l* and half heterozygous. The former gives all *maroon-like* progeny in the backcross with *ma-l* males, and the latter gives all wild-type since the cross is the same as the original cross.

21.14 When females of a particular mutant strain of *Drosophila melanogaster* are crossed to wild-type males, all the viable progeny flies are females. Hypothetically, this result could be the consequence of either a sex-linked, lethal mutation or a maternally inherited factor that is lethal to males. What crosses would you perform in order to distinguish between these alternatives?

Answer: The first possibility is that the results are the consequence of a sex-linked lethal gene. The females would be homozygous for a dominant gene L that is lethal in males but not in females. In this case mating the F_1 females of an L/L x $+/Y$ cross to $+/Y$ males should give a sex ratio of 2 females:1 male in the progeny flies.

The second possibility is that the trait is cytoplasmically transmitted via the egg and is lethal to males. In this case the same F_1 females should continue to have only female progeny when mated with $+/Y$ males.

21.15 Reciprocal crosses between two *Drosophila species, D. melanogaster and D. simulans,* produce the following results:

melanogaster female x *simulans* male → females only
simulans female x *melanogaster* male → males, with few or no females

Propose a possible explanation for these results.

Answer: The simplest explanation is that a *D. simulans* X chromosome is needed for hybrid survival in *D. melanogaster* cytoplasm. However, in *D. simulans* cytoplasm the presence of a *D. melanogaster* X chromosome is generally lethal.

21.16 Some *Drosophila* flies are very sensitive to carbon dioxide; they become anesthetized when it is administered to them. The sensitive flies have a cytoplasmic particle called *sigma* that has many properties of a virus. Resistant flies lack *sigma*. The sensitivity to carbon dioxide shows strictly maternal inheritance. What would be the outcome of the following two crosses: (a) sensitive female x resistant male; and (b) sensitive male x resistant female?

Answer: a. All progeny flies will be sensitive.
b. All progeny flies will be resistant.

21.17 In yeast a haploid nuclear (segregational) *petite* is crossed with a neutral petite. Assuming that both strains have no other abnormal phenotypes, what proportion of the progeny ascospores are expected to be *petite* in phenotype if the diploid zygote undergoes meiosis?

Answer: *1/2 petite,* 1/2 wild-type *(grande)*

21.18 When grown on a medium containing acriflavin, a yeast culture produces a large number of very small *(tiny)* cells that grow very slowly. How would you determine whether the slow growth phenotype was the result of a cytoplasmic factor or a nuclear gene?

Answer: This problem is formally very similar to determining the inheritance mode for yeast *petite* mutants. If the *tiny* phenotypes are due to a nuclear gene, then from a cross of *tiny x normal,* meiotic segregation should generate a 1:1 ratio of *tiny:normal* cells in the progeny (cf. properties of segregational *petites).* If, in contrast, an extranuclear gene is involved, segregation will not be evident, and all progeny will be either *normal* or *tiny.*

21.19 *In Neurospora* a chromosomal gene *F* suppresses the slow-growth characteristic of the *poky* phenotype and makes a *poky* culture into a *fast-poky* culture, which still has abnormal cytochromes. Gene *F* in a combination with normal cytoplasm has no detectable effect. (Hint: Since both nuclear and extranuclear genes have to be considered, it will be convenient to use symbols to distinguish the two. Thus cytoplasmic genes will be designated in square brackets; e.g., [*N*] for normal cytoplasm, [*po*] for *poky.)*
   a. A cross in which *fast-poky* is used as the female (proto-perithecial) parent and a normal wild-type strain is used as male parent gives half *poky* and half *fast-poky* progeny ascospores. What is the genetic interpretation of these results?
   b. What would be the result of the reciprocal cross of the cross described in part a, that is, normal female x *fast-poky* male?

Answer: a. Parents: [*po*]*F* female x [*N*]+ male; progeny: [*po*]*F* and [*po*]+ in equal numbers. Since standard *poky* [*po*]+ is found among the offspring, gene *F* must not effect a permanent alteration of *poky* cytoplasm. The 1:1 ratio of *poky* to *fast-poky* indicates that all progeny have the poky mitochondrial phenotype (by maternal inheritance) and that the *F* gene must be a nuclear gene segregating according to Mendelian principles. Thus the *poky* progeny are [*po*]$^+$ and the fast-poky are [*po*]*F*.

b. Parents: [*N*]+ x [*po*]*F;* progeny: [*N*]$^+$ and [*N*]*F* in equal numbers. These two types are phenotypically indistinguishable.

21.20 Reciprocal crosses between two types of the evening primrose, *Oenothera hookeri* and *Oenothera muricata,* produce the following effects on the plastids:

> *0. hookeri* female x *0. muricata* male → yellow plastids
> *0. muricata* female x *0. hookeri* male → green plastids

Explain the difference between these results, noting that the chromosome constitution is the same in both types.

Answer: The plastids are primarily derived from the egg cytoplasm. There is a difference in the effect of the hybrid *hookeri/muricata* nuclear gene combination on different plastids. When *hookeri* is the maternal parent, the *hookeri* plastids that are maternally inherited become yellow in the *hookeri/muricata* nuclear genetic background. When *muricata* is the maternal parent, the maternally inherited *muricata* plastids remain green in the *hookeri/muricata* nuclear genetic background.

21.21 A form of male sterility in maize is maternally inherited. Plants of a male-sterile line crossed with normal pollen give male-sterile plants. Some lines of maize carry a dominant, so-called restorer (*Rf*) gene, which restores pollen fertility in male-sterile lines.
a. If a male-sterile plant is crossed with pollen from a plant homozygous for gene *Rf*, what will be the genotype and phenotype of the F$_1$?
b. If the F$_1$ plants of part a are used in females in a testcross with pollen from a normal plant.(*rf/rf*), what would be the result? Give genotypes and phenotypes, and designate the type of cytoplasm.

Answer: a. If normal cytoplasm is [*N*] and male-sterile cytoplasm is [*Ms*], then the F$_1$ genotype is [*Ms*]*Rf/rf,* and the phenotype is male-fertile.
b. The cross is [*Ms*]*Rf/rf* female x [*N*]*rf/rf* male giving 50 percent [*Ms*]*Rf/rf* and 50 percent [*Ms*]*rf/rf* progeny. Thus, the phenotypes are 1/2 male fertile and 1/2 male sterile.

21.22 A few years ago the political situation in Chile was such that very many young adults were kidnapped, tortured, and killed by government agents. When abducted young women had young children or were pregnant those children were often taken and given to government supporters to raise as their own. Now that the political situation has changed, grandparents of stolen children are trying to locate and reclaim their grandchildren. Imagine that you are a judge in a trial centering on the custody of a child. Mr. and Mrs. Escobar believe Carlos Mendoza is the son of their abducted, murdered daughter. If this is true, then Mr. and Mrs. Sanchez are the paternal grandparents of the child, as their son (also abducted and murdered) was the husband of the Escobars' daughter. Mr. and Mrs. Mendoza claim Carlos is their natural child. The attorney for the Escobar and Sanchez couples informs you that scientists have discovered a series of RFLPs in human mitochondrial DNA. He tells you his clients are eager to be tested, and ask that you order that Mr. and Mrs. Mendoza and Carlos be tested also.
a. Can mitochondrial RFLP data be helpful in this case? In what way?
b. Do all seven parties need to be tested? If not, who actually needs to be tested in this case? Explain your choices.
c. Assume the critical people have been tested, and you have received the results. How would the results determine your decision?

Answer: a. Carlos Mendoza will have inherited his mitochondrial DNA from his mother, and she will have inherited it from her mother. If Mrs. Mendoza and Mrs. Escobar have different mitochondrial RFLPs, then it can be determined which of them contributed mitochondria to Carlos.

215

b. None of the potential grandfathers need to be tested, since they will not have given any mitochondria to Carlos. In addition, there is no point in testing Mrs. Sanchez. She may have given mitochondria to Carlos' father, but the father did not pass them on to Carlos.

c. If Mrs. Mendoza and Mrs. Escobar do not differ in RFLP, the data will not be helpful. If they do differ and Carlos matches Mrs. Mendoza, the case should be dismissed. If Carlos matches Mrs. Escobar, then the Escobar and Sanchez couples are indeed the grandparents, and the Mendozas have claimed a stolen child.

21.23 The pedigree in the figure below shows a family in which an inherited disease called Leber's optic atrophy is.segregating. This condition causes blindness in adulthood. Studies have recently shown that the mutant gene causing Leber's optic atrophy is located in the mitochondrial genome.

a. Assuming II-4 marries a normal person, what proportion of his offspring should inherit Leber's optic atrophy?

b. What proportion of the sons of II-2 should be affected?

c. What proportion of the daughters of II-2 should be affected?

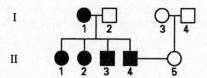

Answer: a. Since II-4 is a male, he will not contribute any of his mutant mitochondria to his offspring. None of them should get Leber's.

b. All the sons should be affected.

c. All the daughters should be affected.

21.24 Imagine you have discovered a new genus of yeast. In the course of your studies on this organism, you isolate DNA and subject it to CsCl density gradient centrifugation. You observe a major peak at a density of 1.75 g/cm^3 and a minor peak at a density of 1.70 g/cm^3 . How could you determine whether the minor peak represents organellar (presumably mitochondrial) DNA, as opposed to a relatively AT-rich repeated sequence in the nuclear genome?

Answer: There are various possibilities. You could isolate the minor peak and examine it using the electron microscope. If the molecules in this peak are circular, they are unlikely to be nuclear fragments. You could grow the yeast in the presence of an intercalating agent such as acridine, and see whether this treatment causes the minor peak species to disappear. If it does, the minor peak is organellar in origin. You could isolate the minor peak, label it (by nick translation, for example), and hybridize this labeled DNA to DNA within suitably prepared yeast cells, and then use the electron microscope to determine whether the label is found over the nucleus or the mitochondria. Finally, you could isolate the minor peak DNA and study its homology to other yeast mitochondria.

CHAPTER 22

QUANTITATIVE GENETICS

I. CHAPTER OUTLINE

The Nature of Continuous Traits
    Why Some Traits Have Continuous Phenotypes
    Questions Studied in Quantitative Genetics
Statistics
    Samples and Populations
    Distributions
    Binomial Theorem
    The Mean
    The Variance and Standard Deviation
    Correlation
    Regression
    Analysis of Variance
Polygenic Inheritance
    Inheritance of Ear Length in Corn
    Polygene Hypothesis For Quantitative Inheritance
    Determining the Number of Polygenes for a Quantitative Trait
Heritability
    Components of the Phenotypic Variance
    Broad-Sense and Narrow-Sense Heritability
    Understanding Heritability
    How Heritability is Calculated
    Identification of Genes Influencing a Quantitative Trait
Response to Selection
    Estimating Response to Selection
    Genetic Correlations
Inbreeding
    Homozygosity and Inbreeding

II. IMPORTANT TERMS AND CONCEPTS

discontinuous trait	multiple gene hypothesis
continuous trait	contributing alleles
quantitative trait	heritability
polygenic trait	broad-sense heritability
multifactorial trait	narrow-sense heritability
familial trait	phenotypic variance
frequency distribution	genetic variance
histogram	environmental variance
normal distribution	genetic-environmental variance
binomial distribution	additive genetic variance
binomial expansion	dominance variance

Pascal's triangle
mean
variance
standard deviation
correlation coefficient
covariance
regression line
regression coefficient
analysis of variance (ANOVA)

interaction variance
quantitative trait locus
selection response
selection differential
phenotypic correlation
genetic correlation
heterosis
inbreeding depression

III. THINKING ANALYTICALLY

Solving the questions and problems given at the end of this chapter puts analytical thinking to the test. Close attention should be paid to the solutions given for the two sample problems, Q1 and Q2, and to problems 22.1, 22.2, 22.3, 22.6, and 22.7. As will be seen in problem 22.7, there is more than one approach to a solution. Devise the one that seems to you to be the most logical and leads most directly to the solution.

IV. QUESTIONS FOR PRACTICE

A. Multiple Choice Questions

1. The ABO blood types represent an example of a
   a. polygenic trait.
   b. continuous trait.
   c. discontinuous trait.
   d. quantitative trait.

2. A trait of which the phenotype is influenced by both multiple genotypes and environmental factors is
   a. multifactorial
   b. polygenic.
   c. discontinuous.
   d. ephemeral.

3. The description of a population in terms of the number of individuals that display varying degrees of expression of a character or range of phenotypes is a
   a. polygraph.
   b. polynomial.
   c. normal distribution
   d. frequency distribution.

4. A symmetrical, bell-shaped curve describes a
   a. binomial distribution.
   b. normal distribution .
   c. frequency histogram.
   d. standard deviation.

5. The general expression for the binomial expansion is
   a. $(p + q)^n$.
   b. $(a^2 + b^2)$
   c. $(a + b)^2$
   d. $a^2 + 2ab + b^2$

6. Variance measures the amount of deviation of a character from the average in a population. This variance is expressed in quantitative terms by the formula

   a. $s = \sqrt{s^2}$

   b. $s^2 = \dfrac{\Sigma(x_i - \bar{x})^2}{n - 1}$

   c. $\dfrac{(x_i - \bar{x})(y_i - \bar{y})}{n - 1}$

   d. $\dfrac{v_{xy}}{s_x s_y}$

7. The coefficients for the binomial expanded to the power $n = 9$ would be (Hint: see Pascal's triangle)
   a. 1:9:32:65:87:65:32:9:1.
   b. 9:36:84:126:84:36:9.
   c. 1:8:28:56:70:56:28:8:1.
   d. 1:9:36:84:126:126:84:36:9:1.

8. The proportion of the phenotypic variance that is attributable to additive genetic variance among individuals in a population defines
   a. heritability.
   b. broad-sense heritability.
   c. narrow-sense heritability.
   d. dominant variance.

9. The proportion of phenotypic variance that consists of genetic variance, additive or otherwise, is called
   a heritability.
   b. broad-sense heritability.
   c. narrow-sense heritability.
   d. environmental variance.

10. The phenomenon called "hybrid vigor" is technically known as
   a. heterosis.
   b. the *cis-trans* effect.
   c. genetic dominance.
   d. genetic codominance.

Answers: 1c, 2a, 3d, 4b, 5a, 6b, 7d, 8c, 9b, 10a

B. Thought Questions

1. Assume that mature fruit weight in pumpkins is genetically determined. Assume also that in the following experiment, environmental factors (weather, soil, etc.) are uniform. Two pumpkin varieties, both of which produce fruit with a mean weight of 20 lbs, are crossed. The F_1 also produces 20-lb. pumpkins. The F_2 plants, however, give the following results:

Mean fruit wt. (lbs)	5	12.5	20	27.5	35
Number of plants	19	82	119	79	21

Explain these results: postulate how many genes are involved and how much each contributes to fruit weight. (Hint: How closely do the data fit a 1:4:6:4:1 phenotypic ratio? What would be the value for n in the binomial $(a + b)^n$? See text p. 685.)

2. Distinguish between heritability, broad-sense heritability, and narrow-sense heritability. How are the latter two put to use by plant and animal breeders? (See text, pp. 687, 689-691.)

3. Skin pigmentation in people is a heritable trait involving several allelic pairs. Can a woman and a man, both of whom have medium color or darkness of skin, have a light-skinned child? A very dark-skinned child? Explain. What degree of darkness would have the greatest probability for a child of theirs? Why?

4. What is meant by genetic correlation, and of what use is it to animal and plant breeders? (See text, 696-698.)

5. Define or describe each of the following: (a) selection differential ; (b) analysis of variance; (c) hybrid vigor; (d) regression analysis. (See text, pp. 681-683, 682-683, 694-696, 698-699.)

V. Answers and Solutions to Text Questions

22.1 The following measurements of head width and wing length were made on a series of steamer-ducks:

Specimen	Head Width (cm)	Wing Length (cm)
1	2.75	30.3
2	3.20	36.2
3	2.86	31.4
4	3.24	35.7
5	3.16	33.4
6	3.32	34.8
7	2.52	27.2
8	4.16	52.7

a. Calculate the mean and standard deviation of head width and of wing length for these eight birds.

b. Calculate the correlation coefficient for the relationship between head width and wing length in this series of ducks.

c. What conclusions can you make about the association between head width and wing length in steamer-ducks?

Answer: a. The mean value is easily obtained: just add up the individual values and divide their sum by the total number of values, hence $25.21/8 = 3.15$ cm is the mean head width. The mean wing length is similarly calculated and equals 35.21 cm. Standard deviation equals the square root of the variance, s^2, which is computed by squaring the sum of the difference between each measurement and the average value and dividing this sum by the number of measurements minus one. Thus you get : $\Sigma(x_i- \overline{x})^2/(n-1) = 1.68/7 = 0.24$ cm^2; from which you extract the square root, which equals 0.49 cm, and this is the value of s, the standard deviation. Repeat this sequence of operations for wing length, and you have the answers to part a: Head width mean = 3.15 cm, and the standard deviation = 0.49 cm. Wing length mean 35.21 cm, and standard deviation 7.68 cm.

b. Correlation coefficient, r, is calculated from the covariance, cov, of two quantities, in this case of head width and wing length, which we may label x and y, respectively, and designate cov as cov_{xy}. This is then divided by the product of the standard deviations of the quantities, x and y. In short, $r = \dfrac{cov_{xy}}{s_x s_y}$. So, start with calculating cov_{xy}, which is expressed by the formula $cov_{xy} = \dfrac{\Sigma x_i y_i - \dfrac{1}{n}(\Sigma x_i \Sigma y_i)}{n-1}$. To get the first factor, you must get the sum of the products of the individual measurements of head width and the corresponding measurements for wing length, which you will find rounds-off to 913 cm. The next factor is the product of the sums of two sets of measurements, divided by the number of the pairs of measurements. This value, $(25.21 \times 284)/8$, rounds off to 887. The difference between these two factors, $913 - 887 = 16$, is then divided by $n - 1$; that is $26/7 = 3.7$. This is the value for cov_{xy}. Finally, the correlation coefficient r can be calculated as $cov_{xy}/s_x s_y$. The values s_x and s_y have already been calculated as 0.49 and 7.68, respectively. By substitution, we get $r = 3.7/3.76 = 0.98$, the answer to part b.

c. Head width and wing length display a strong positive correlation, meaning that ducks with wider heads tend to have longer wings.

22.2   Using methods described in this chapter, answer the following questions.
    a.   In a family of six children, what is the probability that 3 will be girls and 3 will be boys?
    b.   In a family of five children, what is the probability that 1 will be a boy and 4 will be girls?
    c.   What is the probability that in a family of six children, all will be boys?

Answer: a.   Obviously this problem calls for the binomial expansion, $\frac{n!}{s!\,t!}a^s b^t$. First identify each factor of this formula with its corresponding factor in the problem statement. Thus n identifies with the number of children, namely six, so $n = 6$. Let s represent the number of girls, so $s = 3$; and t the number of boys, so $t = 3$ too. Finally, the quantities a and b represent the individual probabilities of a girl and of a boy, respectively. Since the probability that a given child will be a girl is 1/2, and the chance that it will be a boy is also 1/2, then $a = 1/2$ and $b = 1/2$. Now the appropriate substitutions may be made:
$$\frac{n!}{s!\,t!}a^s b^t = \frac{6!}{3!\,3!}\,(1/2)^3(1/2) = \frac{720}{36}\,(1/8)(1/8) = 20(1/64) = 20/64.$$

    b.   Part b is solved in a similar way, but by changing the value of n from 6 to 5, and the values of s and t to 4 for girls and 1 for boys. Hence $\frac{5!}{4!\,1!}\,(1/2)^4(1/2)^1 = 5/32$

    c.   Part c is much more readily solved. If the chance of one boy = 1/2, the chance of six boys = $(1/2)^6$, or 1/64.

22.3   In flipping a coin, there is a 50 percent chance of obtaining heads and a 50 percent chance of obtaining tails on each flip. If you flip a coin 10 times, what is the probability of obtaining exactly 5 heads and 5 tails?

Answer: This is another problem for the binomial expansion. Let n equal the number of flips (10), s the number of heads (5), t the number of tails (5), and a and b the separate probabilities of a head (1/2) or a tail (1/2) respectively. Then $\frac{10!}{5!\,5!}\,(1/2)^5(1/2)^5 = 252/1024$.

22.4   The F_1 generation from a cross of two pure-breeding parents that differ in a size character is usually no more variable than the parents. Explain.

Answer: Each pure-breeding parent is homozygous for the genes (however many there are) controlling the size character, and hence each parent is homogeneous in type. A cross of two pure-breeding strains will generate an F_1 heterozygous for those loci controlling the size trait. Since the F_1 is genetically homogeneous (all heterozygotes), it shows no greater variability than the parents.

22.5   If two pure-breeding strains, differing in a size trait, are crossed, is it possible for F_2 individuals to have phenotypes that are more extreme than either grandparent (i.e., be larger than the largest or smaller than the smallest in the parental generation)? Explain.

Answer: If the two grandparental types (i.e., the two parents in the parental generation) represent the two extremes for the size trait, then the answer is no. If each is within the range, and if each strain differs in the appropriate genes, then it is possible to get an F_2 that is more extreme than the grandparents. For example, consider a quantitative trait in which three pairs of alleles, *Aa, Bb,* and *Cc,* are involved. Suppose further that each capital-letter allele contributes a certain amount to a quantitative character. If, for example, we performed the cross *AA BB cc* x *aa bb CC,* then the F_1 would be *Aa Bb Cc.* The first parental has four capital-letter alleles, and the second has two. Selfing the F_1 results in some F_2 progeny being *AA BB CC* and some being *aa bb cc.* The former has six capital-letter alleles, and the latter has no capital-letter alleles; and thus they will have phenotypes that are more extreme than the phenotypes of either grandparent.

22.6   Two pairs of genes with two alleles each, A/a and B/b, determine plant height additively in a population. The homozygote $AA\ BB$ is 50 cm tall, the homozygote $aa\ bb$ is 30 cm tall.
   a.   What is the F_1 height in a cross between the two homozygous stocks?
   b.   What genotypes in the F_2 will show a height of 40 cm after an F_1 x F_1 cross?
   c.   What will be the F_2 frequency of the 40 cm plants?

Answer: For the purposes of this problem, it is fair to assume that A and B contribute equally to height in excess of the basic height of 30 cm, namely 5 additional centimeters each.
   a.   P: $AA\ BB$ x $aa\ bb$
        F_1: $Aa\ Bb \rightarrow$ two contributing alleles add 10 cm to basic height = 40 cm.
   b.   Of the F_2 1/16 would be $AA\ bb$, 4/16 would be $Aa\ Bb$, and 1/16 $aa\ BB$, each of which would contribute 10 cm to plant height to achieve the total height of 40 cm.
   c.   As shown in Part b, 1/16 $AA\ bb$ + 4/16 $Aa\ Bb$ + 1/16 $aa\ BB$ = 6/16 would be the genotypes of the 40-cm plants of the F_2.

22.7   Three independently segregating genes (A, B, C), each with two alleles, determine height in a plant. Each capital-letter allele adds 2 cm to a base height of 2 cm.
   a.   What are the heights expected in the F_1 progeny of a cross between homozygous strains $AA\ BB\ CC$ (14 cm) x $aa\ bb\ cc$ (2 cm)?
   b.   What is the distribution of heights (frequency and phenotype) expected in an F_1 x F_1 cross?
   c.   What proportion of F_2 plants will have heights equal to the heights of the original two parental strains, $AA\ BB\ CC$ and $aa\ bb\ cc$?
   d.   What proportion of the F_2 will breed true for the height shown by the F_1?

Answer: a.   The F_1 would have the genotype $Aa\ Bb\ Cc$, with 3 contributing alleles and so would contribute 6 cm to the base height of 2 cm for a sum height of 8 cm.
   b.   There are at least three ways of arriving at a solution to this part: one way is by making a diagram like a Punnett square that would show the variety and frequencies of the F_2 genotypes and phenotypes, and accordingly the number of capital letters in each. This diagram would be set up as below, in which the genotypes represent the F_2 distribution of each pair of alleles that comprise the complete F_2, and the numbers represent the number of capital letters in each combination thereof:

	AA	Aa	Aa	aa	
	6	5	5	4	cc
BB	5	4	4	3	Cc
	5	4	4	3	Cc
	4	3	3	2	cc
	5	4	4	3	cc
Bb	4	3	3	2	Cc
	4	3	3	2	Cc
	3	2	2	1	cc
	5	4	4	3	cc
Bb	4	3	3	2	Cc
	4	3	3	2	Cc
	3	2	2	1	cc
	4	3	3	2	cc
bb	3	2	2	1	Cc
	3	2	2	1	Cc
	2	1	1	0	cc

Alternatively, for the less ambitious, there is Pascal's triangle from which the coefficients for $n = 6$ can be had at a glance. Finally, extension of the binomial theorem can be applied, with the same result, which gives the answer to Part b as:
   1 (2 cm):6 (4 cm):15 (6 cm):20 (8 cm):15 (10 cm):6 (12 cm):1 (14 cm).

c. The proportion of *AA BB CC* genotype $(1/4)^3 = 1/64$; similarly the proportion of *aa bb cc* genotype $(1/4)^3 = 1/64$; and the proportion with heights equal to the extremes is $= (1/4)^3 + (1/4)^3 = 2/64$.

d. None of the F_2 with the height of 8 cm can breed true, because this height depends upon a genotype having three capital-letter alleles, which in turn requires heterozygosity for at least one gene locus.

22.8 Repeat Problem 22.7, but assume that each capital-letter allele acts to double the existing height; for example, *Aa bb cc* = 4 cm, *AA bb cc* = 8 cm, *AA Bb cc* = 16 cm, and so on.

Answer:
a. 16 cm
b. 1 (2 cm):6 (4 cm):15 (8 cm):20 (16 cm):15 (32 cm):6 (64 cm):1 (128 cm)
c. 2/64
d. The 16-cm F_1 height is determined by the presence of three dominant genes, meaning that one gene pair must always be heterozygous. Therefore a 16-cm strain could not breed true.

22.9 Assume three equally and additively contributing pairs of genes control flower length in nasturtiums. A completely homozygous plant with 10 mm flowers is crossed to a completely homozygous plant with 30 mm flowers. F_1 plants all have flowers about 20 mm long. F_2 plants show a range of lengths from 10 to 30 mm, with about 1/64 of the F_2 having 10 mm flowers and 1/64 having 30 mm flowers. What distribution of flower length would you expect to see in the offspring of a cross between an F_1 plant and the 30 mm parent?

Answer: The fact that 1/64 of the F_2 reach either parental extreme indicates that 3 loci (6 alleles) controlling flower length are segregating in this cross. Since they all contribute equally, we can conclude that each capital allele contributes 1/6 of the 20 mm difference between the parents, or 3.33 mm. We can designate the genotype of the F_1 as *Aa Bb Cc* and that of the 30 mm parent as *AA BB CC*. The 30 mm parent will contribute *A B C* to every offspring, so the minimum length of the offspring will be 20 mm. We use the binomial theorem to find out that 1/8 of the gametes of the F_1 parent will contain no capital alleles and 1/8 will contain 3 capitals, while 3/8 will contain 1 and 3/8 will contain 2 capitals. Thus we expect 1/8 20 mm, 3/8 23.33 mm, 3/8 26.67 mm, and 1/8 30 mm.

22.10 In a particular experiment, the mean internode length in spikes of the barley variety *asplund* was found to be 2.12 mm. In the variety *abed binder* the mean internode was found to be 3.17 mm. The mean of the F_1 of a cross between the two varieties was approximately 2.7 mm. The F_2 gave a continuous range of variation from one parental extreme to the other. Analysis of the F_2 generation showed that in the F_2 8 out of the total 125 individuals were of the *asplund* type, giving a mean of 2.19 mm. Eight other individuals were similar to the parent *abed binder,* giving a mean internode length of 3.24 mm. Is the internode length in spikes of barley a discontinuous or a quantitative trait? Why?

Answer: It is a quantitative trait. Variation appears to be continuous over a range rather than falling into three discrete classes. Also, the pattern in which the F_1 mean falls between the parental means and in which the F_2 individuals show a continuous range of variation from one parental extreme to the other is typical of quantitative inheritance.

22.11 From the information given in Problem 22.10, determine how many gene pairs involved in the determination of internode length are segregating in the F_2.

Answer: Two allelic pairs are segregating in the F_2. This result is indicated by the fact that 8 of the total 125 individuals, or approximately 1/16, represent each parental extreme. This fraction comes from the binomial expansion of $(a + b)^4$, where $a = b = 1/2$. The coefficients in the expansion are 1:4:6:4:1.

22.12   Assume that the difference between a type of oats yielding about 4 g per plant and a type yielding 10 g is the result of three equal and cumulative multiple-gene pairs *AA BB CC*. If you cross the type yielding 4 g with the type yielding 10 g, what will be the phenotypes of the F_1 and the F_2? What will be their distribution?

Answer: The cross is *aa bb cc* (4 g) x *AA BB CC* (10 g), giving an F_1 that is *Aa Bb Cc* (7 g). The F_2 will show a range of phenotypes from 4 to 10 g, with relative proportions of the seven types being the coefficients in the binomial expansion of $(a + b)^6$ = 1 (4 g):6 (5 g):15 (6 g):20 (7 g):15 (8 g):6 (9 g):1 (10 g).

22.13   Assume that in squashes the difference in fruit weight between a 3-lb type and a 6-lb type is due to three allelic pairs, *A/a, B/b,* and *C/c*. Each capital-letter allele contributes a half pound to the weight of the squash. From a cross of a 3-lb plant (*aa bb cc*) with a 6-lb plant (*AA BB CC*), what will be the phenotypes of the F_1 and the F_2? What will be their distribution?

Answer: The cross is *aa bb cc* (3 lb) x *AA BB CC* (6 lb), which gives an F_1 that is *Aa Bb Cc* and which weighs 4.5 lb. The distribution of phenotypes in the F_2 is, 1 (3 lb):6 (3.5 lb):15 (4 lb):20 (4.5 lb):15 (5 lb):6 (5.5 lb):1 (6 lb).

22.14   Refer to the assumptions stated in Problem 22.13. Determine the range in fruit weight of the offspring in the following squash crosses: (a) *Aa Bb CC* x *aa Bb Cc;* (b) *AA bb Cc* x *Aa BB cc;* (c) *aa BB cc* x *AA BB cc.*

Answer: a. Each capital letter allele contributes 1/2 lb to the baseline weight of 3 lb. From this cross the maximum number of capital-letter alleles a progeny individual can have is five, and the minimum number is one. The former gives 5.5-lb squash, and the latter gives a 3.5-lb squash; so the range is 3.5 to 5.5 lb.
        b. The maximum number of capital-letter alleles possible is four, and the minimum number is two, giving a range of 4 to 5 lb.
        c. All the progeny are *Aa BB cc,* giving 4.5-lb squashes since there are three capital-letter alleles.

22.15   Assume that the difference between a corn plant 20 dm (decimeters) high and one 26 dm high is due to four pairs of equal and cumulative multiple alleles, with the 26-dm plants being *AA BB CC DD* and the 10-dm plants being *aa bb cc dd*.
        a.   What will be the size and genotype of an F_1 from a cross between these two true-breeding types?
        b.   Determine the limits of height variation in the offspring from the following crosses:
             (1) *Aa BB cc dd* x *Aa bb Cc dd;*
             (2) *aa BB cc dd* x *Aa Bb Cc dd;*
             (3) *AA BB Cc DD* x *aa BB cc Dd;*
             (4) *Aa Bb Cc Dd* x *Aa bb Cc Dd.*

Answer: a.   The progeny are *Aa Bb Cc Dd,* which are 18 dm high.
        b.   (1) The minimum number of capital-letter alleles and the maximum number is four, giving a height range of 12 to 18 dm.
             (2) The minimum number is one, and the maximum number is four, giving a height range of 12 to 18 dm.
             (3) The minimum number is four, and the maximum number is six, giving a height range of 18 to 22 dm.
             (4) The minimum number is zero, and the maximum number is seven, giving a height range of 10 to 24 dm.

22.16   Refer to the assumptions given in Problem 22.15. But for this problem two 14-dm corn plants, when crossed, give nothing but 14-dm offspring (case A). Two other 14-dm plants give one 18-dm, four 16-dm, six 14-dm, four 12-dm, and one 10-dm offspring (case B). Two other 14-dm plants, when crossed, give one 16-dm, two 14-dm, and one 12-dm offspring (case C). What genotypes for each of these 14-dm parents (cases A, B, and C) would explain these results? Would it be possible to get a plant taller than 18-dm by selection in any of these families?

Answer: The cross in case A could be between two plants each of which is homozygous for a capital-letter allele at one locus only. Or it could be between a plant homozygous for a capital-letter allele at one locus and a plant homozygous for a capital-letter allele at another locus, such as *AA bb cc dd* x *aa BB cc dd*. Further breeding could distinguish these crosses.

The cross in case B gives proportions that are the coefficients in the binomial expansion of $(a + b)^4$. This result suggests heterozygosity at two loci, such as *Aa Bb cc dd* x *Aa Bb cc dd*.

The cross in case C gives proportions that are the coefficients in the binomial expansion of $(a + b)^2$. Possible genotypes are *AA bb cc dd* and *Aa Bb cc dd*.

Plants with heights greater than 18 dm would require five capital-letter alleles. Since in no cases are more than two gene loci involved in the particular crosses described, then any individual can have a maximum of only four capital-letter alleles, so it would not be possible to select plants taller than 18 dm.

22.17 A quantitative geneticist determines the following variance components for leaf width in a population of wild flowers growing along a roadside in Kentucky.

Additive genetic variance (V_A)	= 4.2
Dominance genetic variance (V_D)	= 1.6
Interaction genetic variance (V_I)	= 0.3
Environmental variance (V_E)	= 2.7
Genetic-environmental variance (V_{GE})	= 0.0

    a. Calculate the broad-sense heritability and the narrow-sense heritability for leaf width in this population of wild flowers.

    b. What do the heritabilities obtained in part a indicate about the genetic nature of leaf width variation in this plant?

Answer: a. Broad-sense heritability =
$$V_G/V_P = (V_A + V_D + V_I)/(V_A + V_D + V_I + V_E + V_{GE})$$
$$= (4.2 + 1.6 + 0.3)/(4.2 + 1.6 + 0.3 + 2.7 + 0)$$
$$= 6.1/8.8 = 0.69$$
Narrow-sense heritability = $V_A/V_P = 4.2/8.8 = 0.48$

    b. About 69 percent of the phenotypic variation in leaf width observed in this population is due to genetic differences among individuals. About 48 percent of the phenotypic variation is due to additive genetic variation.

22.18 Assume all genetic variance affecting seed weight in beans is genetically determined and is additive. From a population where the mean seed weight was 0.88 g, a farmer selected two seeds, each weighing 1.02 g. He planted these and crossed the resulting plants to each other. then collected and weighed their seeds. The mean weight of their seeds was 0.96 g. What is the narrow-sense heritability of seed weight?

Answer: The selection differential is 0.14 g. The selection response is 0.08 g. Therefore h^2 is 0.08/0.14 = 0.57.

22.19 Members of the inbred rat strain SHR are salt sensitive: they respond to high salt environment by developing hypertension. Members of a different inbred rat strain, TIS, are not salt sensitive. Imagine you placed a population consisting only of SHR rats in an environment that was variable in regard to distribution of salt, so that some rats would be exposed to more salt than others. What would be the heritability of blood pressure in this population?

Answer: All the rats with SHR genes would have high blood pressure, while all those with TIS genes would have low blood pressure.

225

22.20   On his farm in Kansas a farmer is growing a variety of wheat called TK138. He calculates the narrow-sense heritability for yield (the amount of wheat produced per acre) and finds that the heritability of yield for TK138 is 0.95. The next year he visits a farm in Poland and observes that a Russian variety of wheat, UG334, growing there has only about 40% as much yield as TK138 grown on his farm in Kansas. Since he found the heritability of yield in his wheat to be very high, he concludes that the American variety of wheat (TK138) is genetically superior to the Russian variety (UG334), and he tells the Polish farmers that they can increase their yield by using TK138. What is wrong with his conclusion?

Answer: Heritability is specific to a particular population and to a specific environment and cannot be used to draw conclusions about the basis of populational differences. Because the environment of the farms in Kansas and in Poland differ, and because the two wheat varieties differ in their genetic makeup, the heritability of yield calculated in Kansas cannot be applied to the wheat grown in Poland. Furthermore, the yield of TK138 would most likely be different in Poland, and might even be less than the yield of the Russian variety when grown under in Poland.

22.21   Dermatoglyphics are the patterns of the ridged skin found on the fingertips, toes, palms, and soles. (Fingerprints are dermatoglyphics.) Classification of dermatoglyphics is frequently based on the number of triradii; a triradius is a point from which three ridge systems separate at angles of 120°. The number of triradii on all ten fingers was counted for each member of several families and the results are tabulated below.

Family	Mean Number Of Triradii In The Parents	Mean Number Of Triradii In The Offspring
I	14.5	12.5
II	8.5	10.0
III	13.5	12.5
IV	9.0	7.0
V	10.0	9.0
VI	9.5	9.5
VII	11.5	11.0
VIII	9.5	9.5
IX	15.0	17.5
X	10.0	10.0

   a.   Calculate the narrow-sense heritability for the number of triradii by the regression of the mean phenotype of the parents against the mean phenotype of the offspring.
   b.   What does your calculated heritability value indicate about the relative contributions of genetic variation and environmental variation to the differences observed in number of triradii?

Answer: a.   The narrow-sense heritability of number of triradii will equal the slope of the regression line of the mean offspring phenotype on the mean parental phenotype. For the data presented in this problem, the slope = 1.03.
           b.   All the observed variation in phenotype can be attributed to additive genetic variation among the individuals.

22.22   A scientist wishes to determine the narrow-sense heritability of tail length in mice. He measures tail length among the mice of a population and finds a mean tail length of 9.7 cm. He then selects the ten mice in the population with the longest tails; mean tail length in these selected mice is 14.3 cm. He interbreeds the mice with the long tails and examines tail length in their progeny. The mean tail length in the F_1 progeny of the selected mice is 13 cm.
   Calculate the selection differential, the response to selection, and the narrow-sense heritability for tail length in these mice.

Answer: Selection differential = 14.3 - 9.7 = 4.6
Selection response = 13 - 9.7 = 3.3
Narrow-sense heritability = selection response/selection differential = 3.3/4.6 = 0.72

22.23   Suppose that the narrow-sense heritability of wool length in a breed of sheep is 0.92, and the narrow-sense heritability of body size is 0.87. The genetic correlation between wool length and body size is −0.84. If a breeder selects for sheep with longer wool, what will be the most likely effect on wool length and on body size?

Answer: Because the heritabilities for body size and wool length are high, these traits should respond to selection. Because a negative correlation exists between body size and wool length, when sheep with longer wool are selected for, body size will decrease.

22.24   The heights of 10 college age males and the heights of their fathers are presented below.

Height of Son (inches)	Height of Father (inches)
70	70
72	76
71	72
64	70
66	70
70	68
74	78
70	74
73	69

    a.   Calculate the mean and the variance of height for the sons and do the same for the fathers.
    b.   Calculate the correlation coefficient for the relationship between the height of father and height of son.
    c.   Determine the narrow-sense heritability of height in this group by regression of the son's height on the height of father.

Answer: a. Fathers: Mean = 71.9, variance = 11.6
          Sons: Mean = 70, variance = 10.25
      b. Correlation = 0.49
      c. Slope = b = 0.46
          Narrow-sense heritability = 2(b) = 2(0.46) = 0.92

22.25   The narrow-sense heritability of egg weight in a particular flock of chickens is 0.60. A farmer selects for increased egg weight in this flock. The difference in the mean egg weight of the unselected chickens and the selected chickens is 10 g. How much should egg weight increase in the offspring of the selected chickens?

Answer: Selection response = narrow-sense heritability x selection differential
Selection response = 0.60 x 10g = 6 g

22.26   Would a law banning marriages between individuals and their stepparents be founded on genetic principles?

Answer: No, since stepparents and stepchildren are usually not close biological relatives. So there is not likely to be as significant a problem of deleterious effects in their offspring as there is among children of parents who are closely related.

22.27   Is there any circumstance for which inbreeding does not have deleterious consequences?

Answer: If there are no deleterious recessives in the inbreeding population, then inbreeding should not have deleterious consequences. This situation is seen in many plant groups in which inbreeding is the rule, since by now most deleterious recessives have been eliminated by selection. (They arise anew, of course, by mutation.)

CHAPTER 23

POPULATION GENETICS

I. CHAPTER OUTLINE

II. IMPORTANT TERMS AND CONCEPTS

Mendelian population	overdominance
gene pool	founder effect
genotypic frequency	bottleneck effect
allelic/gene frequency	adaptation
genetic drift	fitness
random mating	selection coefficient
Hardy-Weinberg equilibrium	assortive mating

balance model *vs.* classical model of genetic
 variation
neutral mutation hypothesis

outbreeding *vs.* inbreeding
multigene family
concerted evolution/molecular drive

III. THINKING ANALYTICALLY

When combined with the other aspects of the overall science of genetics, classical population genetics requires a shift of focus away from a consideration of individual molecular mechanisms of specific biochemical processes to a statistical evaluation of the effects of those processes at the level of the group, population, or species. The analytical tools are those of statistics applied to the phenomena underlying the Hardy-Weinberg equilibrium as well as deviations there from.

Recently, the powerful techniques of molecular genetics have been trained on some of the evolutionary questions implicit in population genetics, with the result that a synthesis of sorts is emerging that encompasses not only the intellectual framework of population biology but also the modes of analytic thought employed to solve problems. The material at the end of the chapter, including such topics as DNA sequence variation, evolution of mitochondrial DNA, and the derivation of relationships by means of nucleic acid sequences and their changes, addresses these issues.

When studying the material in this chapter, it is important to spend the time necessary to understand the meaning of the various equations that are used to measure changes in populations. Simply memorizing the equations takes less time, but is foolhardy because it is much more difficult, if not impossible, to solve even moderately difficult word problems by just plugging numbers into equations. Although it is helpful to recognize by sight several of the equations, that will not allow completion of the chapter-end problems.

One final caution - allelic symbolism in population genetics is often different than for Mendelian or molecular genetics. Whereas the genotypes for normal and sickle-cell hemoglobin are written as $\beta^A\beta^A$, $\beta^A\beta^S$, $\beta^S\beta^S$ in Chapter 8, here they are are written as *Hb-A/Hb-A*, *Hb-A/Hb-S*, *Hb-S/Hb-S*. Similarly, care should be taken with the letters *p* and *q*, which symbolize the frequency within a population of dominant and recessive members of an allelic pair, with *P*, *H*, and *Q*, which represent genotypic frequencies of the dominant homozygote, heterozygote, and recessive heterozygote, respectively.

As you solve the practice problems below and the chapter end problems in the text, do not forget that the frequency of recessive homozygotes (q^2) is often the only piece of hard data available. As you work back from that information keep track of the assumptions that you are making so that you do not enter the realm of circular reasoning. For example, if you calculate *q* from its square root, obtain *p* by subtracting *q* from 1, and then determine that heterozygotes exist at a frequency equal to 2*pq*, you are <u>assuming</u> that random mating is occurring. Further work with your calculated values of p^2, 2*pq*, and q^2 will continue to reflect that assumption. More information is required (or, in some cases, statistical derivations that are beyond the scope of your text) in order to use those values to prove that the population in question is at Hardy-Weinberg equilibrium.

IV. QUESTIONS FOR PRACTICE

A. Multiple Choice Questions

1. In a small population, 30% is of blood group M, 40% is of blood group MN, and 30% is of blood group N. What are the gene frequencies for the two alleles in question? (Let *p* = the frequency of the L^M allele, and *q* = the frequency of the L^N allele.)
  a. *p* = 0.3; *q* = 0.7
  b. *p* = 0.5; *q* = 0.5
  c. *p* = 0.3; *q* = 0.3
  d. *p* = 0.5; *q* = 0.3

2. Is the population described in #1, above, in Hardy-Weinberg equilibrium?
  a. Yes, because the calculated genotypic frequencies equal the expected genotypic frequencies.
  b. Yes, because at equilibrium you would always have equal numbers of recessive and dominant homozygotes.
  c. No, the frequency of heterozygotes is too large.
  d. No, the frequency of heterozygotes is too low.

3. In a population of North American whites, about 8% of the males are red/green color blind. Assuming the population is in equilibrium, what proportion of the females would you expect to be similarly affected?
    a. 0.64%
    b. 4%
    c. 64%
    d. 0.92%

4. In a large, randomly mating population, 80% of the individuals have dark hair and 20% are blond. Assuming that hair color is controlled by one pair of alleles, is the allele for dark hair <u>necessarily</u> dominant to the one for blond hair?
    a. Yes, because otherwise the population would not be dominated by dark-haired individuals.
    b. Yes, because more of something (in this case, color) is always dominant to less of that thing.
    c. No, because relative frequencies of alleles in a randomly mating population is unrelated to issues of dominance and recessiveness.
    d. No, because although there is a relationship between dominance and allele frequency, that relationship is not seen in this example.

5. Which of the following populations is in genotypic equilibrium?

	Genotypes		
Population	AA	Aa	aa
a	0.72	0.20	0.08
b	0.12	0.80	0.08
c	0.25	0.50	0.25
d	0.08	0.01	0.91

6. Would you expect the rate of forward mutation to be higher than or lower than the rate of back mutation?
    a. The rate of forward mutation is generally lower because there is mutational pull back to a specific form.
    b. The rate of forward mutation is generally higher because once an allele has changed, it is just as likely that a subsequent change will be to yet another new form.
    c. The rate of forward and back mutations are co-dependent and therefor are usually equal.
    d. The relative rates of forward and back mutation are so highly variable that one can not formulate an accurate generalization comparing the two.

7. What is the effective population size for a population consisting of 10 breeding males and 2 breeding females?
    a. 12
    b. 6
    c. 7
    d. 2

8. Genetic drift is caused by
    a. the founding of a population by a very small number of "pioneers".
    b. any traumatic event that markedly reduces the size of a population, such as being hunted to near-extinction.
    c. environmental factors that continuously and effectively limit the number of surviving members of a group.
    d. all of the preceding can cause genetic drift.

9. Migration and genetic drift:
   a. have similar effects on the size and variability within populations.
   b. both tend to decrease both population size and variability.
   c. both tend to increase both population size and variability.
   d. have opposite effects on size and variability. Migration effectively increases size and variability, whereas drift acts in opposition.

10. In which sections of DNA would you expect to find the highest rate of mutation:
    a. non-functional pseudogenes.
    b. introns.
    c. exons.
    d. leaders and trailers.

Answers: 1b, 2d, 3a, 4c, 5c, 6b, 7c, 8d, 9d, 10a

B. Thought Questions

1. Show why equilibrium values for genotypic frequencies are reached in one generation after the onset of random mating for autosomal alleles, but more than one generation of random mating is required for sex-linked alleles. (See text, p. 712.)
2. Distinguish the concept of heterozygosity with that of proportion of polymorphic loci. (See text pp. 711-712, 720.)
3. Explain why mutation pressure alone is rarely the most important determinant of gene frequency. (See text, pp. 726-735.)
4. What is the most likely explanation if the gene frequencies agree with those predicted by the Hardy-Weinberg Law, but the genotype frequencies do not? (Hint: think about mating systems and also heterozygote superiority).
5. Explain why mutation, migration, and drift do not necessarily lead to adaptation. (See text pp. 746-747.)

V. ANSWERS AND SOLUTIONS TO TEXT QUESTIONS

23.1   In the European land snail, *Cepaea nemoralis*, multiple alleles at a single locus determine shell color. The allele for brown (C^B) is dominant to the allele for pink (C^P) and to the allele for yellow (C^Y). Pink is recessive to brown, but is dominant to yellow, and yellow is recessive to pink and brown. Thus, the dominance hierarchy among these alleles is $C^B > C^P > C^Y$.

In one population of *Cepaea*, the following color phenotypes were recorded:

Brown	236
Pink	231
Yellow	33
Total	500

Assuming that this population is in Hardy-Weinberg equilibrium (large, randomly mating, and free from outside evolutionary forces), calculate the frequencies of the C^B, C^Y, and C^P alleles.

Answer:

Brown $= C^B C^B, C^B C^P, C^B C^Y = p^2 + 2pq + 2pr = 236/500 = 0.472$

Pink $= C^P C^P, C^P C^Y, = q^2 + 2qr = 231/500 = 0.462$

Yellow $= C^Y C^Y = r^2 = 33/500 = 0.066$

$f(C^Y C^Y) = r^2$

$\sqrt{f(c^Y c^Y)} = qr^2$

$\sqrt{33/500} = r = \sqrt{0.066} = 0.26$

$f(C^P C^P + C^P C^Y + C^Y C^Y) = q^2 = 2qr + r^2$

$f(C^P C^P + C^P C^Y + C^Y C^Y) = (q + r)^2$

232

$$\frac{231 + 33}{500} = (q + r)^2$$

$$0.528 = (q + r)^2$$

$$\sqrt{0.528} = (q + r)^2$$

$$0.727 = q + r$$

$$r = 0.26$$

$$0.727 - r = q$$

$$q = 0.467$$

$$p = 1 - q - r$$

$$= 1 - 0.467 - 0.26 = 0.273$$

So, $f(C^B) = p = 0.273$

$f(C^Y) = r = 0.26$

$f(C^P) = q = 0.467$

23.2   Three alleles are found at a locus coding for malate dehydrogenase (MDH) in the spotted chorus frog. Chorus frogs were collected from a breeding pond and each frog's genotype at the MDH locus determined with electrophoresis. The following numbers of genotypes were found:

M^1M^1	8
M^1M^2	35
M^2M^2	20
M^1M^3	53
M^2M^1	76
M^3M^3	62
Total	254

a.   Calculate the frequencies of the M^1, M^2, and M^3 alleles in this population.

b.   Using a chi-square test, determine whether the MDH genotypes in this population are in Hardy-Weinberg proportions.

Solution: a.   The tally for M^1 is as follows:

genotype	# individuals	# M^1 alleles
M^1M^1	8	16
M^1M^2	35	35
M^1M^3	53	53
	Total	104

The total number of individuals in the population is 254, thus the total number of alleles equals 254 x 2, or 508. Hence, the frequency of the M^1 allele is 104/503, which equals 0.20   The frequencies of the M^2 and M^3 alleles is similarly obtained:

$$f(M^1) = 0.20 = p$$
$$f(M^2) = 0.30 = q$$
$$f(M^3) = 0.50 = r$$

b.   Since this system consists of three alleles, their relationship is predicted by the following expression:

$$p^2 + 2pq + 2pr + q^2 + 2qr + r^2 = 1$$

To obtain the expected number in a population of 254 individuals, the frequency of each genotype must be multiplied by 254. For example, $p = 0.04$. Hence 0.04(254), or 10 individuals would be expected in this population.

	M^1M^1	M^1M^2	M^1M^3	M^2M^2	M^2M^3	M^3M^3
	p^2	$2pq$	$2pr$	q^2	$2qr$	r^2
Expected Number (e)	10	30	51	23	76	64
Observed Number (o)	8	35	53	20	76	62
Deviation (d)	-2	+5	+2	-3	0	-2
d^2	4	25	4	9	0	4
d^2/e	0.4	0.83	0.08	0.39	0	0.06

$\chi^2 = 1.76$

The number of phenotypes (6) minus the number of alleles (3) determines the degrees of freedom (3). With 3 degrees of freedom, $P > 0.05$. Hence, the deviations appear to be caused by chance alone, and this population is in Hardy-Weinberg equilibrium.

Answer: a. $f(M^1) = 0.20$
$f(M^2) = 0.30$
$f(M^3) = 0.50$

              b. $\chi^2 = 1.76$, degrees of freedom $= 6 - 3 = 3$, $P > 0.05$. This population appears to be in Hardy-Weinberg proportions.

23.3   In a large interbreeding population 81 percent of the individuals are homozygous for a recessive character. In the absence of mutation or selection, what percentage of the next generation would be homozygous recessives? Homozygous dominants? Heterozygotes?

Answer: Let p = probability of A, and q = probability of a. Then $q^2 = 0.81$, $q = 0.9$, and $p = 0.1$. Therefore we expect, in the next generation, 0.81 aa, 0.01 AA, and $(2)(0.9)(0.1) = 0.18$ Aa.

23.4   Let A and a represent dominant and recessive alleles whose respective frequencies are p and q in a given interbreeding population at equilibrium (with $p + q = 1$).
        a. If 16 percent of the individuals in the population have recessive phenotypes, what percentage of the total number of recessive genes exist in the heterozygous condition?
        b. If 1.0 percent of the individuals were homozygous recessive, what percentage of the recessive genes would occur in heterozygotes?

Answer: a. $\sqrt{0.16} = 0.4 = 40\%$ = frequency of recessive alleles; $1 - 0.4 = 0.6 = 60\%$ = frequency of dominant alleles; $2pq = (2)(0.4)(0.6) = 0.48$ = probability of heterozygous diploids. Then $(0.48)/[(2 \times 0.16) + 0.48] = 0.48/0.80 = 60\%$ of recessive alleles are in heterozygotes.
            b. If $q^2 = 1\% = 0.01$, then $q = 0.1$, $p = 0.9$, and $2pq = 0.18$ heterozygous diploids. Therefore $(0.18)/[(2 \times 0.01) + 0.18) = 0.18/0.20 = 90\%$ of recessive alleles in heterozygotes.

23.5   A population has eight times as many heterozygotes as homozygous recessives. What is the frequency of the recessive gene?

Answer: Since $2pq$ = number of heterozygotes and q recessive homozygotes, if there are eight times as many heterozygotes as recessive homozygotes then:

$2pq = 8q^2$
Dividing both sides by q^2: $2pq/q^2 = 8$
Hence: $2p/q = 8$
And: $2p = 8q$
But since $p+q = 1$; $p = 1 - q$
Substituting for p: $2(1-q) = 8q$
$2 - 2q = 8q$
$2 = 10q$
$q = 0.2$ = the frequency of the recessive allele

23.6 In a large population of range cattle the following ratios are observed: 49 percent red (*RR*), 42 percent roan (*Rr*), and 9 percent white (*rr*).
  a. What percentage of the gametes that give rise to the next generation of cattle in this population will contain allele *R*?
  b. In another cattle population only 1 percent of the animals are white and 99 percent are either red or roan. What is the percentage of *r* alleles in this case?

Answer: a. 49% + 1/2(42%) = 70%
            b. $\sqrt{0.01} = 0.1 = 10\%$

23.7 In a population gene pool the alleles *A* and *a* have initial frequencies of *p* and *q*, respectively. Prove that the gene frequencies and zygotic frequencies do not change from generation to generation as long as there is no selection, mutation, or migration, the population is large, and the individuals mate at random.

Answer: The zygotic frequencies generated by random mating are p^2 *AA* + 2*pq Aa* + q^2 *aa* = 1. All of the gametes of the *AA* individuals and half the gametes of heterozygotes will bear the *A* allele. Then the frequency of *A* in the gene pool of the next generation is $p^2 + pq = p(p + q) = p$. Thus each generation of random mating under Hardy-Weinberg conditions fails to change either the allelic or zygotic frequencies. The more lengthy proof of this result, considering all possible matings, their frequencies, and the relative frequencies of different genotypes in their offspring, is given in the text.

23.8 The *S-s* antigen system in humans is controlled by two codominant alleles, *S* and *s*. In a group of 3146 individuals the following genotypic frequencies were found: 188 *SS*, 717 *Ss*, and 2241 *SS*.
  a. Calculate the frequency of the *S* and *s* alleles.
  b. Determine whether the genotypic frequencies conform to the Hardy-Weinberg equilibrium by using the chi-square test.

Answer: a. Let *p* = the frequency of *S* and *q* = the frequency of *s*. Then

$$p = \frac{2(188)\ SS + 717\ Ss}{2(3146)} = \frac{1093}{6292} = 0.1737$$

$$q = \frac{717\ Ss + 2\ (2241)\ ss}{2(3146)} = \frac{5199}{6292} = 0.8263$$

  b.

Class	Observed	Expected	*d*	d^2/e
SS	188	94.9	+93.1	91.235
Ss	717	903.1	-186.1	38.361
ss	2241	2148.0	+93.0	4.032
	3146	3146.0	0	133.628

There is only one degree of freedom because the three genotypic classes are completely specified by two gene frequencies, namely, *p* and *q*. Thus the number of degrees of freedom number of genes −1. The χ^2 value for this example is 133.628, which, for one degree of freedom, gives a *P* value less than 0.0001. Therefore the distribution of genotypes differs significantly from the Hardy-Weinberg equilibrium.

23.9 Refer to Problem 23.8. A third allele is sometimes found at the *S* locus. This allele S^u is recessive to both the *S* and the *s* alleles and can only be detected in the homozygous state. If the frequencies of the alleles *S*, *s*, and S^u are *p*, *q*, and *r*, respectively, what would be the expected frequencies of the phenotype *S−, Ss, s−,* and $S^u S^u$?

Answer: frequency of $S^- = p^2 + 2pr$, frequency of $Ss = 2pq$; frequency of $s^- = q^2 + 2qr$; frequency of $s^u s^u = r^2$.

23.10 In a large interbreeding human population 60 percent of individuals belong to blood group O (genotype $I^O I^O$). Assuming negligible mutation and no selective advantage of one blood type over another, what percentage of the grandchildren of the present population will be type O?

Answer: Under the conditions described the population is in equilibrium, so the answer is 60%.

23.11 A selectively neutral, recessive character reappears in 0.40 of the males and in 0.16 of the females in a randomly interbreeding population. What is the gene's frequency? How many females are heterozygous for it? How many males are heterozygous for it?

Answer: Since the frequency of the trait is different in males and females, the character might be caused by a sex-linked recessive gene. If the frequency of this gene is q, females would occur with the character at a frequency of q^2, and males with the character would occur at a frequency of q. The frequency of males is $q = 0.4$, and thus we may predict that the frequency of females would be $(0.4)^2 = 0.16$ if this is a sex-linked gene. This result fits the observed data. Therefore the frequency of heterozygous females is $2pq = 2(0.6)(0.4) = 0.48$. For sex-linked genes no heterozygous males exist.

23.12 Suppose you found two distinguishable types of individuals in wild populations of some organism in the following frequencies:

	Type 1	Type 2
Females	99%	1%
Males	90%	10%

The difference is known to be inherited. What is its genetic basis?

Answer: A sex-linked pair of alleles, occurring with frequencies of 0.1 recessive and 0.9 dominant.

23.13 Red-green color blindness is due to a sex-linked recessive gene. About 64 women out of 10,000 are color-blind. What proportion of men would be expected to show the trait if mating at random?

Answer: $0.0064 = q^2$ so $q = 0.08$ = probability of color-blind male

23.14 About 8 percent of men in a population are red-green color-blind (owing to a sex-linked recessive gene). Answer the following questions, assuming random mating in the population, with respect to color blindness.
   a. What percentage of women would be expected to be color-blind?
   b. What percentage of women would be expected to be heterozygous?
   c. What percentage of men would be expected to have normal vision two generations later?

Answer: Let p = frequency of recessive for color blindness, $(c) = 0.08$ and let q = frequency of dominant $(C) = (1 - p) = 0.92$.
   a. Frequency of color-blind women is given by $p^2 = (0.08)^2 = 0/064$; i.e., 0.64% of women are color-blind.
   b. Frequency of heterozygous women is given by $2pq = (2)(0.08)(0.92) = 0.1473$; i.e., 14.72% of women are heterozygotes.
   c. Frequencies of alleles will not change in two generations given random mating in a population in Hardy-Weinberg equilibrium. Thus frequency of men with normal vision two generations later is given by $q = 0.92$, i.e., 92% of men will have normal vision in two generations.

23.15 List some of the basic differences in the classical, balance, and neutral-mutation models of genetic variation.

Answer: Classical Model – Little genetic variation within populations; one allele in high frequency (wild type); most individuals homozygous for wild-type; low variation maintained by strong selection for the wild-type allele.

Balance Model – Much genetic variation within populations; many alleles in intermediate frequencies; many heterozygotes; variation maintained by balancing selection.

Neutral-Mutation Model – Much variation; many alleles in intermediate frequencies; many heterozygotes; variation neutral with regards to natural selection.

23.16 Two alleles of a locus, A and a, can be interconverted by mutation:

$$A \underset{v}{\overset{u}{\rightleftarrows}} a$$

and u is a mutation rate of 6.0×10^{-7} and v is a mutation rate of 6.0×10^{-8}. What will be the frequencies of A and a at mutational equilibrium, assuming no selective difference, no migration, and no random fluctuation caused by genetic drift?

Answer:

$$q = \frac{u}{u+v} = \frac{6 \times 10^{-7}}{(6 \times 10^{-7}) + (6 \times 10^{-8})} = \frac{6 \times 10^{-7}}{(6 \times 10^{-7}) + (0.6 \times 10^{-7})} = \frac{6}{6.6} = 0.91$$

$$p = 1 - q = 1 - 0.91 = 0.09$$

Thus, the frequencies are 0.008 AA, 0.16 Aa, and 0.828 aa.

23.17

a. Calculate the effective population size (N_e) for a breeding population of 50 adult males and 50 adult females.

b. Calculate the effective population size (N_e) for a breeding population of 60 adult males and 40 adult females.

c. Calculate the effective population size (N_e) for a breeding population of 10 adult males and 90 adult females.

d. Calculate the effective population size (N_e) for a breeding population of 2 adult males and 98 adult females.

Answer: The effective size of a breeding population is described by the equation:

$$N_e = \frac{4 \times N_f \times N_m}{N_f + N_m}$$

Hence:
a. $(4 \times 50 \times 50)/100 = 100$
b. $(4 \times 60 \times 40)/100 = 96$
c. $(4 \times 10 \times 90)/100 = 36$
d. $(4 \times 2 \times 98)/100 = 7.8$

23.18 In a population of 40 adult males and 40 adult females, the frequency of allele A is 0.6 and the frequency of allele a is 0.4.

a. Calculate the 95% confidence limits of the gene frequency for A.

b. Another population with the same gene frequencies consists of only 4 adult males and 4 adult females. Calculate the 95% confidence limits of the gene frequency for A in this population.

c. What are the 95% confidence limits of A if the population consists of 76 female and 4 males?

Answer: a. $N_e = \dfrac{4(N_m)(N_f)}{(N_m + N_f)} = \dfrac{4(40)(40)}{(40 + 40)} = 80$

$$s_p = \sqrt{pq/2N_e} = \sqrt{(0.6)(0.4)/2(80)} = \sqrt{0.0015} = 0.039$$

95% confidence limits $= p \pm 2s_p = 0.522 \le p \le 0.844$

b. $N_e = 8$

$$s_p = \sqrt{pq/2N_e} = \sqrt{(0.6)(0.4)/2(8)} = \sqrt{0.015} = 0.122$$

95% confidence limits $= p \pm 2s_p = 0.356 \le p \le 0.844$

c. $N_e = 15.2$

$$s_p = \sqrt{pq/2N_e} = \sqrt{(0.6)(0.4)/2(15.2)} = \sqrt{0.00789} = 0.089$$

95% confidence limits $= p \pm 2s_p = 0.422 \le p \le 0.778$

23.19 The land snail *Cepaea nemoralis* is native to Europe but has been accidently introduced into North America at several localities. These introductions occurred when a few snails were inadvertently transported on plants, building supplies, soil, or other cargo. The snails subsequently multiplied and established large, viable populations in North America.

Assume that today the average size of *Cepaea* populations found in North America is equal to the average size of *Cepaea* populations in Europe. What predictions can you make about the amounts of genetic variation present in European and North American populations of *Cepaea*? Explain your reasoning.

Answer: Since the North American populations were founded by a relatively few number of individuals, we would predict that genetic drift has influenced the North American populations to a greater degree than the European populations. Therefore, we might expect to see less variation within and greater genetic differentiation among the North American populations.

23.20 A population of 80 adult squirrels reside on campus, and the frequency of the *Est*[1] allele among these squirrels is 0.70. Another population of squirrels is found in a nearby woods, and there, the frequency of *Est*[1] allele is 0.5. During a severe winter, 20 of the squirrels from the woods population migrate to campus in search of food and join the campus population. What will be the gene frequency of *Est*[1] in the campus population after migration?

Answer: $p'_{II} = mp_I + (1 - m)p_{II}$
$= (0.20)(0.5) + (1 - 0.20)(0.70) = 0.66$

23.21 Upon sampling three populations and determining genotypes, you find the following three genotype distributions. What would each of these distributions imply with regard to selective advantages of population structure?

Population	AA	Aa	aa
1	0.04	0.32	0.64
2	0.12	0.87	0.01
3	0.45	0.10	0.45

Answer: For population 1, out of every hundred individuals, 4 have the genotype *AA*, 32 are *Aa*, and 64 are of genotype *aa*. One hundred individuals contain 200 alleles, hence:

	# *A* alleles	# *a* alleles
4 *AA* individuals	8	0
32 *Aa* individuals	32	32
64 *aa* individuals	0	128
Totals	40	160

238

Thus the frequency of A is $40/200 = 0.2$ and the frequency of a is $160/200 = 0.8$. In a similar way, the frequencies of A and a for each of populations 2 and 3 are: $A = 0.555$, $a = 0.445$ for population 2, and $A = 0.5$, $a = 0.5$ for population 3. Reference to text Table 23.12 shows that population 1 selects against the dominant allele, population 2 selects for the heterozygote, and population 3 selects against the heterozygote.

23.22 The frequency of two adaptively neutral alleles in a large population is 70 percent A:30 percent a. The population is wiped out by an epidemic, leaving only four individuals, who produce many offspring. What is the probability that the population several years later will be 100 percent AA? (Assume no mutations.)

Answer: 100% AA would be expected if all four individuals were homozygous dominant. The probability of this result occurring is $(0.7)^2$ to the fourth power, or $(0.49)^4$ (which is about $(1/2)^4$, or 1/16). The result of 100% AA might be achieved even if not all four survivors were homozygous dominant and the precise answer would be a probability somewhat greater than $(0.49)^4$.

23.23 A completely recessive gene, owing to changed environmental circumstances, becomes lethal in a certain population. It was previously neutral, and its frequency was 0.5.
   a. What was the genotype distribution when the recessive genotype was not selected against?
   b. What will be the gene frequency after one generation in the altered environment?
   c. What will be the gene frequency after two generations?

Answer:
   a. When selectively neutral, the genes distribute themselves binomially, so 0.25 are AA, 0.05 are Aa, and 0.25 are aa.
   b. $q = 0.33$
   c. $q = 0.25$

23.24 Human individuals homozygous for a certain recessive autosomal gene die before reaching reproductive age. In spite of this removal of all affected individuals, there is no indication that homozygotes occur less frequently in succeeding generations. To what might you attribute the constant rate of appearance of recessives?

Answer: mutation of A to a

23.25 A completely recessive gene (Q^1) has a frequency of 0.7 in a large population, and the Q^1Q^1 homozygote has a relative fitness of 0.6.
   a. What will be the frequency of Q^1 after one generation of selection?
   b. If there is no dominance at this locus (the fitness of the heterozygote is intermediate to the fitnesses of the homozygotes), what will the gene frequency be after one generation of selection?
   c. If Q^1 is dominant, what will the gene frequency be after one generation of selection?

Answer:
   a. $\Delta q = \dfrac{-spq^2}{1 - sq^2} = [((-0.4)(0.3)(0.7)^2)/(1 - (0.4)(0.7)^2)] = -0.07$
   $q' = 0.63$

   b. $\Delta q = \dfrac{-spq^2}{1 - sq^2} = [((-0.4)(0.3)(0.7)^2)/(1 - (0.4)(0.7))] = -0.058$
   $q' = 0.7 - 0.058 = 0.64$

   c. To use the formula given in Table 23.12 for selection against a dominant allele we let $p =$ the frequency of the dominant allele selected against and calculate as follows:
   $\Delta p = \dfrac{-spq^2}{1 - s + sq^2} = [((-0.4)(0.7)(0.3)^2)/(1 - 0.4 + (0.4)(0.3)^2)] = -0.040$
   $q' = 0.7 - 0.04 = 0.66$

23.26 As discussed earlier in this chapter, the gene for sickle-cell anemia exhibits overdominance. The *Hb-A/Hb-S* heterozygote has increased resistance to malaria and therefore has greater fitness than the *Hb-A/Hb-A* homozygote, who is susceptible to malaria, and the *Hb-S/Hb-S* homozygote, who has sickle-cell anemia. Suppose that the fitness values of the genotypes in Africa are as presented below:

$$Hb\text{-}A/Hb\text{-}A = 0.88$$
$$Hb\text{-}A/Hb\text{-}S = 1.00$$
$$Hb\text{-}S/Hb\text{-}S = 0.14$$

Give the expected equilibrium frequencies of the sickle-cell gene (*Hb-S*).

Answer: $q = s/(s + t) = 0.12/(0.12 + 0.86) = 0.12$

23.27 Achondroplasia, a type of dwarfism in humans, is caused by an autosomal dominant gene. The mutation rate for achondroplasia is about 5×10^{-5} and the fitness of achondroplastic dwarfs has been estimated to be about 0.2, compared to unaffected individuals. What is the equilibrium frequency of the achondroplasia gene based on this mutation rate and fitness value?

Answer: $q = u/s = (5 \times 10^{-5})/0.8 = 0.0000625$

23.28 The frequencies of the *M* and *N* blood group alleles are the same in each of the populations I, II, and III, but the genotypes' frequencies are not the same, as shown below. Which of the populations is most likely to show each of the following characteristics: random mating, inbreeding, genetic drift. Explain your answers.

	L^M/L^N	L^M/L^N	L^N/L^N
I	0.50	0.40	0.10
II	0.49	0.42	0.09
III	0.45	0.50	0.05

Answer: In all three populations the allele frequencies are $f(L^M) = 0.7$, $f(L^N) = 0.3$. Having said that, it is easy to identify population II as showing the expected Hardy-Weinberg genotype frequencies. Population I shows an excess of homozygotes and population III shows an excess of heterozygotes. Inbreeding increases the frequency of homozygotes for any given allele frequencies, so population I must be the inbred one. Genetic drift, since it represents random changes from Hardy-Weinberg expectations due to small population size, could be associated with any kind of departure from expectations. Population III could be explained by drift, but not by inbreeding.

23.29 DNA was collected from 100 people randomly sampled from a given human population and was digested with the restriction enzyme *Bam*HI, electrophoresed, and Southern blotted. The blots were probed with a particular cloned sequence. Three different patterns of hybridization were seen on the blots. Some DNA samples (56 of them) showed a single band of 6.3 kb, others (6) showed a single band at 4.1 kb, and yet others (38) showed both the 6.3 and the 4.1 kb bands.
  a. Interpret these results in terms of *Bam*HI sites.
  b. What are the frequencies of the restriction site alleles?
  c. Does this population appear to be in Hardy-Weinberg equilibrium for the relevant restriction site(s)?

Answer: a. The data fit the idea that a single *Bam*HI site varies. The probe is homologous to a region wholly within the 4.1 kb piece bounded on one end by the variable *Bam*HI site, and on the other end by a constant site. When the variable site is present, the hybridized fragment is 4.1 kb. When the variable site is absent, the fragment extends to the next constant *Bam*HI site, and is 6.7 kb long. People with only 4.1 or only 6.7 kb bands are homozygotes, people with both are heterozygotes.
  b. The "+" allele of the variable site is present in $2(6) + 38 = 50$ chromosomes, and the "-" allele is present in $2(56) + 38 = 150$ chromosomes. Thus $f(+)$ is 0.25 and $f(-)$ is 0.75.

c. If the population is in Hardy-Weinberg equilibrium, we would expect $(0.25)^2$, or 0.0625 of the sample to show only the 4.1 kb band. This would be 6.25 individuals. We observed 6. We expect $(0.75)^2$ or 0.5625 to be homozygous for the 6.7 kb band, which is 56.25 individuals. We saw 56. Finally, we would expect $2(025 \times 0.75)$ or 0.375 to be heterozygotes, or 37.5 individuals. We observed 38. The observed numbers are so close to the expected that a χ^2 test is unnecessary.

23.30 DNA was isolated from 10 nine-banded armadillos and cut with the restriction enzyme *Hin*dIII. *Hin*dIII recognizes the six base sequence $\begin{smallmatrix}5'\text{-AAGCTT-}3'\\3'\text{-TTCGAA-}5'\end{smallmatrix}$. The DNA fragments that resulted from the restriction reaction were separated with agarose electrophoresis and transferred to nitrocellulose using Southern blotting. A labeled probe for the β-hemoglobin gene was added and the following set of restriction patterns was observed, where +/+ indicates that the restriction site was present on both chromosomes of the individual, +/- indicates that the restriction site was present on one chromosome and absent on one chromosomes of the individual, and -/- indicates that the restriction site was absent on both chromosomes of the individual:

Calculate the expected heterozygosity in nucleotide sequence.

Answer: To calculate the expected heterozygosity in nucleotide sequence, we use the formula:

$$H_{nuc} = \frac{n(\Sigma c_i) - \Sigma c_i^2}{j(\Sigma c_i)(n-1)}.$$

In this equation, n equals the number of homologous DNA molecules examined. In our example we looked at 10 armadillos, each with two homologous chromosomes, so $n = 20$. The quantity j equals the number of nucleotides in the restriction site; in our example this is six, because *Hin*dIII recognizes a six-base sequence. For each restriction site i, c_i represents the number of molecules in the sample that were cleaved at that restriction site. Since we examined only a single restriction site, and that site was cut in 10 of the 20 chromosomes, $\Sigma c_i = 10$. Thus, we obtain

$$H_{nuc} = \frac{20(10) - 10^2}{6(10)(19)} = \frac{200 - 100}{1140} = 0.088$$

23.31 Fifty tiger salamanders from one pond in west Texas were examined for genetic variation using the technique of protein electrophoresis. Each salamander was genotyped for five loci (*AmPep, ADH, PGM, MDH,* and *LDH*-1). No variation was found at *AmPep, ADH,* and *LDH*-1; in other words, all individuals were homozygous for the same allele at these loci. The following numbers of genotypes were observed at the *MDH* and *PGM* loci.

MDH Genotypes	Number of Individuals
A A	11
A B	35
B B	4

PGM Genotypes	Number of Individuals
DD	35
DE	10
EE	5

Calculate the number of polymorphic loci and the heterozygosity for this population.

Answer: To calculate the percent polymorphic loci we divide the number of loci with more than one allele (2) by the total number of loci examined (5), i.e. 2/5 = 0.40. Heterozygosity is calculated by averaging the frequency of heterozygotes for each locus. The frequency of heterozygotes for the *AmPep*, *ADH*, and *LDH*-1 loci is zero. 35 out of 50 individuals are heterozygous at the *MDH* locus and 10 out of 50 individuals are heterozygous at the *PGM* locus. Thus,

$$\frac{0 + 0 + 0 + 35/50 + 10/50}{5} = \frac{0.9}{5} = 0.18.$$

23.32 What factors cause genetic drift?

Answer: Continuous small population size, small number of founders (founder effect), and a reduction in population size (bottleneck effect).

23.33 What are the primary effects of the following evolutionary forces on the gene and genotypic frequencies of a population?
   a. mutation.
   b. migration.
   c. genetic drift.
   d. inbreeding.

Answer:
   a. Leads to change in gene frequencies within a population if no other forces are acting. Introduces new genetic variation. If population size is small, mutation may lead to genetic differentiation among populations.
   b. Increases population size and increases genetic variation within populations. Equalizes gene frequencies among populations.
   c. Reduces genetic variation within populations and leads to genetic change over time. Increases genetic differences among populations. Increases homozygosity within populations.
   d. Increases homozygosity within populations. Decreases genetic variation.

23.34 Explain how overdominance leads to an increased frequency of sickle-cell anemia in areas where malaria is widespread.

Answer: Individuals heterozygous for the sickle cell gene (*Hb-A/Hb-S*) are at a selective advantage because the abnormal hemoglobin mixture in these individuals provides an unfavorable environment for the growth of malarial parasites. Thus, heterozygotes have higher fitness than *Hb-A/Hb-A* homozygotes who are susceptible to malaria. Heterozygotes also have higher fitness than *Hb-S/Hb-S* homozygotes who suffer from sickle cell anemia. Because the sickle cell gene (*Hb-S*) is favored in the heterozygote, it has relatively high frequency in areas with malaria.

23.35 Suppose we examine the rates of nucleotide substitution in two 300-nucleotide sequences of DNA isolated from humans. In the first sequence (sequence A), we find a nucleotide substitution rate of 4.88×10^{-9} substitutions per site per year. The substitution rate is the same for synonymous and nonsynonymous substitutions. In the second sequence (sequence B), we find a synonymous substitution rate of 4.66×10^{-9} substitutions per site per year and a nonsynonymous substitution rate of 0.70×10^{-9} substitutions per site per year. Referring to Table 23.14, what might you conclude about the possible functions of sequence A and sequence B?

Answer: Comparison with the rates of nucleotide substitution in Table 23.14 indicates that sequence A has as high a rate as that typically observed in mammalian pseudogenes. In addition, the rates of synonymous and nonsynonymous substitutions are the same. These observations suggest that sequence A is either a pseudogene or is a sequence that provides no function, since high rates are substitution are observed when sequences are functionless. Sequence B has a relatively low rate of nonsynonymous substitution but a relatively high rate of synonymous substitution. This is the pattern we expect when a sequence codes for a protein; thus, sequence B probably encodes a protein.

23.36   What are some of the characteristics of mitochondrial DNA evolution in animals?

Answer: More rapid rate of evolution than that observed in nuclear DNA.

23.37   What is concerted evolution?

Answer: Concerted evolution is the maintenance of sequence uniformity in multiple copies of a gene. It leads to similar sequences in multiple copies of the same gene within a species but different sequences among different species.

23.38   What are some of the advantages of using DNA sequences for inferring evolutionary relationships?

Answer: DNA sequences reflect direct genetic differences, they are easily quantified, and all organisms have them.